CW01510139

Pacific Voices and Climate Change

Niki J. P. Alsford
Editor

Pacific Voices and Climate Change

palgrave
macmillan

Editor
Niki J. P. Alsford
University of Central Lancashire
Preston, UK

ISBN 978-3-030-98459-5   ISBN 978-3-030-98460-1   (eBook)
https://doi.org/10.1007/978-3-030-98460-1

This Palgrave Macmillan imprint is published by the registered company Springer Nature Switzerland AG
The registered company address is: Gewerbestrasse 11, 6330 Cham, Switzerland

This book is dedicated to all those affected by the eruption of Hunga Tonga-Hunga Ha'apai underwater volcano and following tsunami that began on 14 January 2022.

CONTENTS

CONTRIBUTORS

Aarushi University of Delhi, Delhi, India

Niki J. P. Alsford University of Central Lancashire, Preston, UK

Betty Barkha Centre for Gender, Peace and Security, Monash University, Melbourne, VIC, Australia

Fanny Caron IrAsia, Aix-Marseille University-CNRS, Marseille, France

Ti-han Chang University of Central Lancashire, Preston, UK

Lyn Collie Auckland, New Zealand

Zakia Firdaus Center for Comparative Literature and Translation Studies, Central University of Gujarat, Gandhinagar, India

Dean Karalekas Centre of Austronesian Studies, University of Central Lancashire, Taipei, Taiwan

Pavan Kumar Janki Devi Memorial College, University of Delhi, Delhi, India

Kate Martin University of Central Lancashire, Preston, UK

Sojin Lim University of Central Lancashire, Preston, UK

Tobie Openshaw Centre of Austronesian Studies, University of Central Lancashire, Taipei, Taiwan

Bob Walley   University of Central Lancashire, Preston, UK

Amar Wayal Department of English, SRM Institute of Science and Technology (Deemed to be University), Tiruchirappalli, India

LIST OF FIGURES

Sustainable Development from Unsustainable Climate: Sustainable Development Goals and the Pacific Small Island Developing States

LIST OF TABLES

**Sustainable Development from Unsustainable Climate:
Sustainable Development Goals and the Pacific Small
Island Developing States**

Introduction

Niki J. P. Alsford

Climate change represents one of humanity's greatest threats. The vastness of the Pacific means that no two experiences are the same. Occupying more than one-third of the Earth's surface, the Pacific Basin, from its fringing rim, its islands, and ocean, is a region that exhibits a unique geographical expanse. It is, for the most part, ocean peppered with islands. It is surrounded by a continental rim that is mostly marked by mountain ranges that trend parallel to the coast. These ranges act as barriers to the movement of air and various living things, including people. This has historically given the Pacific Basin an unmatched climatic and biotic integrity (Nunn 2007: 17). The principal control over climate within this basin is solar radiation that is distributed by latitude. The regional regulator on climate and ocean circulation is the configuration of its differing land and sea systems. As such, the main determinant of environmental change in the region is the alteration of this balance. Environments within the Pacific have changed for multifarious reasons over the last millennia. Understanding this change has not always been simple.

N. J. P. Alsford (✉)
University of Central Lancashire, Preston, UK
e-mail: njpalsford@uclan.ac.uk

N. J. P. Alsford (ed.), *Pacific Voices and Climate Change*, https://doi.org/10.1007/978-3-030-98460-1_1

1

Scientific understanding of how the climate has changed over the last hundred years, for which there are instrumental records and climate proxies, is the least controversial period. Changes occurring within these years witnessed far less variation, until now. Today the principal process driving regionwide environmental change is undoubtedly sea-level rise. Patrick Nunn (2007) argues that, historically, there were three significant periods of environmentally determined change within the Pacific. The first, termed the 'Medieval Warm Period' (750–1250AD), witnessed a stable climate. Here archaeological records show a flourishing of cultural exchange and the beginnings of complex irrigation systems boosting agricultural activity. Warmer waters meant diverse and extensive marine productivity. It is within this period that long-distance voyaging across the Pacific occurred. Migration, therefore, within this period took place during a period of relative climate stability as opposed to one of forced displacement. Around 1250–1350, a significant climatic event, the origin of which has multiple proposed causes, occurred that saw a rapid drop in temperatures and sea levels. This event witnessed a sharp depletion in resources that resulted in food shortages and significant societal change. It was following this event that we witnessed the erection of megalithic monuments, such as the Moai on Rapa Nui (Easter Island) and Nan Madol on Pohnpei in Micronesia. On the island of Tonga, evidence is clear in the popularity of statue production. The event in question led to a Little Ice Age (1350–1800) that saw continued rapid climate change. The Pacific Ocean witnessed an increased number of high-intensity storms, large-scale population dispersal, changes in resource utilisation, and a reduction in interconnected trade interaction. The period witnessed a rise in regional conflict and a cessation of long-distance voyaging. In certain locations, coastal communities abandoned lowland settlements for higher ground, and we witness the occupation of smaller islands, and indication of sporadic displacement. The recent warming within the Pacific has given rise to unprecedented complexities in Pacific societies. The colonisation of the Pacific in the nineteenth and twentieth centuries and the confirming of legal international boundaries has meant that movements of people, as a response to periods of climate change, are no longer possible. What is more, climatic change in the Pacific today is anthropogenic. This has meant that understanding this challenge warrants new analyses.

Worldview expressions by the peoples who live within this rim are embedded within cultural traditions that form part of the peoples' heritage and identity. Passed down through generations, specific belief systems act as guardians of traditional knowledge. Usually referred to as TK, this includes various types, such as subsistence knowledge, or techniques that are used in agricultural practice, fishing, and hunting; ecological knowledge that includes ethnobotany; and health and well-being, which incorporates traditional medicinal practice, social health, and spiritual balance. Although 'tradition' is used to conceptually describe the aforementioned knowledge systems, it is important to note that this is not stagnant and inert. On the contrary, it is continually being amended and built upon. TK is a vital asset. It belongs to its people. It reflects their identity, their community's history, their values, and most importantly, it is used as indicators in their understanding of the surrounding environment. These reflections foster a relational approach between the human and non-human worlds, and this association between peoples and their environment provides a useful setting in understanding and documenting changes within the natural environment of the Pacific.

The multidisciplinary nature and the combination of both academic and non-academic writing have broadened the intended target audience. Attention to this is important as strains on already fragile systems have only been exacerbated by the ongoing COVID-19 pandemic. The reduced media attention on climate change in the wake of the pandemic has shifted the world's attention. This book is an attempt to maintain space for continued discussions on climate change in the Pacific. It is a timely analysis of the impacts of climate change on Pacific coastlines, communities, and societies. Using current research to document climate change via contextually informed research, the authors engage with local cultures, histories, knowledges, and communities. The rates of change within our climate are progressing fast and the peoples of the Pacific region are among those who contributed least to this change, but are among the worst affected. This book examines the problems of environmental change on traditional life and culture from a transdisciplinary context. The chapters that make up this volume were contributed by both academic and non-academic writers.

The first chapter, by Bob Walley, explores the key challenges facing Pacific Island communities because of changes in climate, particularly rising sea levels and storm surges. The chapter concentrates on the Kiribati islands and looks at the debates surrounding the modern displacements of its peoples. The chapter opens the discussion of indigenous perspective, knowledge, and experience in any future migration and adaption initiatives. This is then followed by Sojin Lim, whose important work explores the problem of adaptation to climate crises in island regions from the perspective of the recent Sustainable Development Goals (SDGs). Lim's chapter examines how climate change among Small Island Developing States (SIDS) is addressed in UN SDG implementation policies. With a focus on the Pacific, Lim argues that while there is an extensive amount of research about SIDS and climate change, few have given particular attention to the SDGs and donor engagement in SIDS. The chapter, using Pacific SIDS, reflects on how unsustainable situations, caused by climate change, are driven in national development strategies and further asks whether the donor community integrates the community's needs, as expressed by the recipients, into their development aid policies.

The third chapter by Ti-han Chang and Lyn Collie looks at New Zealand's political responses to climate change and migration in the Pacific. This chapter examines the history of immigration in New Zealand from the British colonial period to contemporary Pacific migrants. The chapter draws attention to census data, revealing more about the social outcomes and lived experiences of Pacific migrants in New Zealand. Using the experiences of I-Kiribati, Chang and Collie argue that I-Kiribati migrants frequently encounter social, cultural, and economic discrimination. The authors review the current Pacific migration schemes offered by the New Zealand government and assess their suitability in managing climate-displaced migration. Continuing the theme of migration, Betty Barkha follows this chapter by exploring the issues surrounding agency and action. Barkha undertakes a critical discourse analysis to closely examine the extent to which gender has been integrated in existing frameworks on climate change-induced human mobility in Fiji.

The exercise in the discourses of migrant interaction forms a sense of entanglement. It is a type of research that concerns itself with linkages and flows of peoples, cultures, and commodities. Pacific island identity has an entangled understanding of regional connectivity due to linguistical and historical periods of migration. The next chapter by Fanny Caron explores indigenous identity in the face of climate change through the works of two young Paiwan authors in Taiwan. Caron-Scarulli argues that, by focussing on climate change, the authors contribute to a literature that opens a path to indigenous imaginary and subsequent indigenous identity. Storytelling and Taiwan are continued in the following chapter by Dean Karalekas and Tobie Openshaw, who argue that the epistemology of Traditional Ecological Knowledge (TEK), which has developed over thousands of years, directly links belief systems to experiential relationships with the natural world and are retold via myths and other narratives. Karalekas and Openshaw draw on the myth-making systems of the Bunun nation in Taiwan and look at how these systems have contributed to a worldview that puts the sentient experience in balance with the ecosystem. Often this worldview is articulated in literary expression as seen in Caron-Scarulli's chapter. The influences of colonial layering interact in how this system of knowledge is disseminated. At first, the following chapter may seem an anomaly; It does not exactly 'fit' the criteria of Pacific indigenous worldviews. Yet, its contribution rests in how the environment plays a pivotal role in indigenous literary landscapes. For Zakia Firdaus and Amar Wayal, the experience of Native American peoples' sense of identity and their connection to the land has led to a conscious effort by Native American nations to express their concerns about environmental degradation and sustainability through engagement with ecocriticism and issues surrounding protection and preservation of natural resources. Firdaus and Wayal's chapter offers a useful opening for a wider comparison to the Pacific Islander experience.

The focus on indigenous Pacific island experience has often led to an image of ethnic homogeneity. The chapter by Kate Martin that follows undertakes an analysis of the experience of Indo-Fijians who share no linguistic, cultural, or historical links to the wider Austronesian-speaking Pacific but offer TEK based on their own lived experience.

The final chapter is more explicit and fosters an integrated reflection on policy. By documenting the similarities and differences between the climate policies of Australia and New Zealand, Aarushi and Pavan Kumar provide not only an important analysis of the strategic implications of climate change policy, but also critically reflect on both governments' responses to climate change incidents.

The discussion of the climate's impact on Pacific Island communities warrants greater attention, not least because no two experiences are the same. Papua New Guinea is very different from Tuvalu. GDP wealth is higher in Hawaii than it is in the Kiribati islands. It is clear that a one-size-fits-all approach does not apply in these circumstances. Instead, what is needed is a set of policies that incorporates traditional knowledge as a critical indicator in its assessment. Current climate change policy recommendations, such as those proffered by the IPCC, often take a 'big picture' approach to how climate change is impacting the eco-systems of the Pacific. Yet, this is just one of many 'big pictures'.

It is a narrow, monocultural way of understanding change. The seas, for example, support the livelihoods of numerous peoples with diverse cultural practices. The best protection against sea-level rise is to engage with a diversity of understandings and responses to this change. Traditional knowledge delivers depth and is an indicator of place-specific cultural contexts. Policy needs to prioritise affected communities by being more culturally inclusive. Co-existence of plurality of knowledges will improve our understanding of complex systems and enable baseline monitoring systems to work effectively in managing human resilience to climate change.

References

IPCC (2021), AR6 Climate Change 2021, accessed 17 January 2022, https://www.ipcc.ch/report/ar6/wg1/#SPM

Nunn, Patrick (2007), *Climate, Environment and Society in the Pacific During the Last Millennium.* Oxford: Elsevier.

The Next Wave of Climate Refugees? Building a Clear Narrative Concerning Levels of Understanding and Agency in Communities Across the Pacific Who Are Most at Risk from the Effects of the Climate Emergency

Bob Walley

This chapter identifies key challenges facing Pacific Island Countries (PICs) caused by the serious impacts of the current climate emergency. Looking at the literature and existing responses to this issue, this chapter will explore issues and debates surrounding the displacement of people from these islands, identifying problems associated with understanding the affected communities' perspectives and their sense of agency and empowerment to do anything about this intensifying threat. This enquiry will focus mainly on Kiribati, a sovereign island nation which gained independence from the United Kingdom in 1979. Kiribati comprises

B. Walley (✉)
University of Central Lancashire, Preston, UK
e-mail: RVWalley1@uclan.ac.uk

© The Author(s), under exclusive license to Springer Nature Switzerland AG 2022
N. J. P. Alsford (ed.), *Pacific Voices and Climate Change*,
https://doi.org/10.1007/978-3-030-98460-1_2

7

of one raised coral island and 32 low-lying atolls, making it increasingly susceptible to climatic impacts such as extreme weather and ocean level rise. Arguments made will highlight the importance of including indigenous perspectives, knowledge and experience in any future mitigation and adaption initiatives. The responsibility and ethical obligation of large Greenhouse Gas emitting countries, especially those with colonial ties to these territories, will be highlighted. The chapter concludes with recommendations for related future research and project work going forwards.

RATIONALE

United Nations Intergovernmental Panel on Climate Change (IPCC) reports present unprecedented threats that PICs face from the escalating climate emergency, including rising sea levels, drought and an increase in strength and frequency of deadly storms (IPCC, 2019a). Pacific island communities are particularly vulnerable to, and suffering from, the adverse effects of climate change related extreme weather effects like coastal erosion and water shortages. The predicted rise in sea levels, altered precipitation patterns, higher temperatures, acidification of the ocean, loss of coastal infra-structure and land, more intense cyclones and droughts, salinisation and crop failure (SPREP, 2015) will exacerbate these risks in the coming decades. This jeopardises the livelihoods of Pacific Island peoples, most of whom are engaged in agriculture, forestry and fishing and are thus dependent on natural resources, whilst also having a particularly detrimental impact on tourism (IPCC, 2019a). Having a 'high ratio of shoreline to land, low elevation, settlement patterns concentrated in coastal areas and a narrow economic basis' (Ferris et al., 2011) puts PICs at extremely high risk. The accelerating climate emergency threatens not only the coastal flora and fauna but their human community's way of life, unique cultural practices and national identities.

THE SITUATION IN KIRIBATI

The Pacific is the least carbon producing human-populated area on earth (Kupferberg, 2021), with PICs being responsible for only an estimated 0.03% of total Green House Gas emissions (GHGe) (IPCC, 2001). Yet a 1.1-m sea-level rise is predicted by 2100 by conservative estimates (IPCC, 2019b) though research shows a 2-m rise is very possible with a 2 °C scenario (Bamber et al., 2019). Even with just 1.1 m of sea-level rise, two-thirds of Kiribati, which has an average height above sea level of 1.8 m (COP23, 2017) could be underwater, with significant parts of its territories becoming uninhabitable. With two meters of sea-level rise, Kiribati would be mostly submerged (Ray, 2019). This makes the prospect of large groups of peoples on PICs being forcibly displaced and seeking to migrate an increasingly likely scenario. For almost a decade, the previous president of Kiribati sought to bring the desperate plight of the country before the international community, pressurising developed nations to take more radical measures at reducing GHGe (Kiribati Office of Climate Change, 2009; Ahmed, 2009). The severity of the IPCC's climate forecasts for many low-lying PICs like Kiribati has created the prospect of 'disappearing states' and prompted debates regarding possible 'statelessness' for many Pacific Islanders (McAdam, 2012). Islanders living on low-lying atolls in the South Pacific have been categorised as some of the world's first 'climate change refugees' (Burch, 2020). However, after initiating an 'experimental humanitarian visa' policy, Aotearoa New Zealand dropped the program because Pacific Islanders rejected it as they did not want to be labelled refugees or have to leave their homes (Ionesco, 2019; Manch, 2018; Dempster & Ober, 2020). This suggests the proud and deep-rooted values which Pacific Islanders hold over their ancestral lands. It's argued 'they prefer support for adaptation and mitigation for their continuing lives on ancestral land with refugee status as merely a ·last resort' (Robie, 2020; Burch, 2020). Therefore, the term 'climate refugee', which is in the very title of this chapter, should be considered carefully and conscientiously.

Only recently did the UN High Commission on Refugees acknowledge any legal basis for climate-related protections (Su, 2020; UNHCR, 2020; Beeler, 2018). Those at risk of imminent harm of losing their 'Right to Life' through climate emergencies can now seek asylum. However, the notion of climate refugee status can create 'a narrow and biased debate' (Ionesco, 2019; Farbotko & Lazarus, 2012). Determining and capturing the proper and most appropriate terminology for migrants of an environmental character is strenuous due to overlapping drivers of migration, such as economic, social and political factors which themselves 'affect migration' (UK Gov, 2011). But why is this terminology important? Besides being an essential element in forming and informing public discourse and opinion both globally and in sending and receiving countries, it is a crucial and invaluable component of drafting effective policies and humanitarian responses on national and international levels (Kupferberg, 2021). Over the last decade, climate-related disasters have become the strongest driver of migration worldwide with an estimated 20 million people a year forced to leave their homes (Oxfam, 2019; Aburn & Wesselbaum, 2017). Whilst the term 'environmental refugee' is currently a highly controversial label, other varieties include climate migrants, ecomigrants, environmental migrants, climate change-induced migrants, ecological refugees, environmental refugees, climate change migrants and environmentally induced forced migrants (Gemenne, 2009). Recent waves of refugee movements in the Middle East, Africa and into Europe show the urgency and need for pre-emptive planning for any future large-scale migrations and movements of people, which are expected to only rise in the decades to come (Yates et al., 2021; Krajick, 2018).

Response Possibilities

There is broad agreement that effective responses to the climate emergency involve both mitigation and adaptation (IPCC, 2014), though the urgency of climate impacts in the Pacific region calls for radical policy interventions (Kupferberg, 2021). 'Temaiku Land and Urban Development' aims to increase the height of a 300-hectare area of swampy inhabitable land on Kiribati's Temaiku Bight, transforming it into an urban development around two meters above predicted 2200 ocean levels (Watkin et al., 2019). The project is expected to take 30 years to complete with land reclamation alone estimated to cost US$273 million (Walters, 2019), more than the entire annual GDP of Kiribati, which was US$194 million in 2019 (World Bank, 2019). As neither of the project leaders, Aotearoa New Zealand or Kiribati, are able to fund the project in full, Kiribati is looking to the Green Climate Fund, a financial mechanism under the UNFCCC and the World Resources Institute's 'Adaptation Finance Accountability Initiative' (Walters, 2019). Nothing like this has ever been successfully carried out. But if successful, the project would house an approximate 35,000 people (Watkin et al., 2019) by 2050, or about a third of Kiribati's current population. However, the population of Kiribati is projected to be more than 239,000 by 2100 (UN, 2019), which puts the long-term viability of this adaptation strategy into serious question (Kupferberg, 2021). Considering current UN population projections, Kiribati would need just under seven Temaiku projects to be able to adapt successfully.

Cross-border planned relocation has almost exclusively happened in the Pacific (Kupferberg, 2021). In Kiribati, the 2014 'Migration With Dignity' (MWD) initiative by former Kiribati president Anote Tong planned to gradually relocate people to Fiji (Pashley, 2016). Two years later Tong was replaced by Taneti Maamau and this plan was scrapped, replaced with a faith-based approach focusing on economic growth, adaptation, mitigation and national pride being adopted (Walker, 2017). Kiribati's status as one of the poorest in the region (Webb, 2020) combined with the amount of investment needed to sufficiently adapt and secure a safe and dignified life for the entire PIC's population only contributes to the strategy's limitations (Kupferberg, 2021). Whilst president Maamau was elected on his conservative vision for Kiribati involving the Christian faith, he does acknowledge the long-term threat of climate change (Walker, 2017). The land purchased in Fiji for MWD is currently farmland, but Maamau has tried to encourage movement out of Kiribati's highly dense and populated capital, where poverty is widespread (Sinha, 2020). However, water inundation and salinisation of the soil are destroying water sources and making land unsuitable for agricultural production (Ferris et al., 2011), disrupting farmers' livelihoods and jeopardising potential financial incentives from the government's economic policy. With 'dark biblical irony' (Kupferberg, 2021), Maamau argues that Kiribati's citizens must 'try to isolate [themselves] from the belief that Kiribati will be drowned [as] the ultimate decision is God's' (Walker, 2017). So whilst the Kiribati president doesn't claim to deny climate change, his religious stance on the impacts of the climate emergency on the country's survival is concerning (Kupferberg, 2021). Furthermore, some citizens of Kiribati (I-Kiribati) and Tuvaluans believe reports of climate change risks are exaggerated or unreliable and God will protect their homelands (Yates et al., 2021; Siose, 2017; Thompson, 2015; Roman, 2013; Gillard & Dyson, 2012).

Nevertheless, people have begun to migrate from their island homes to Aotearoa New Zealand, Australia, the United States or Fiji, expecting better education and employment opportunities, improved health and quality housing. However, many are then shocked by the realities of life abroad (Yates et al., 2021). I-Kiribati, Marshallese and Tuvaluans encountered many challenges and barriers to resettlement, which have left them unemployed, underemployed, in inadequate housing or unable to access education and healthcare services. Resettlement has also often been hindered by strict visa requirements, non-transferral of education credits, workplace exploitation, costly housing, unfamiliar foods; cultural misunderstandings, discomfort speaking English or racism (McClain et al., 2020; Maekawa et al., 2019; McClain et al., 2019; Drinkall et al., 2019; Emont & Anandarajah, 2017; Siose, 2017; Thompson, 2015; Malua, 2014; Roman, 2013; Shen & Binns, 2012; Gillard & Dyson, 2012; Shen & Gemenne, 2011; Gemenne, 2010). Xenophobia especially can 'thwart political action aimed at increasing responsibility sharing and better coordination', particularly in large-scale scenarios (Miller, 2018). Pacific Islanders who relocated to other PICs have often been met by hostile environments in their new destinations, which has been due in part to the difficulty in ensuring land entitlements (Connell, 2007).

Additionally, in Aotearoa New Zealand, some migrants are saddened that their lifestyles are less communal than in their homelands, with less-frequent gatherings, weaker community-based culture and closed-off housing where they feel separated from their neighbours (Yates et al., 2021; Siose 2017; Thompson, 2015; Gillard & Dyson, 2012; Gemenne, 2010). Emphasis on rigid time schedules, monetization and individualism experienced in Pacific Rim countries have proven disruptive for those accustomed to the values of collectivism, respect and self-sufficiency which are common in Kiribati, Tuvalu and the Republic of the Marshall Islands (Yates et al., 2021; McClain et al., 2020; Drinkall et al., 2019; Siose, 2017; Thompson, 2015; Roman, 2013; Locke, 2009). Even the most carefully planned relocation can carry such significant psychosocial and cultural costs that many people only consider it a last resort (Initiative, 2015). Climate-related migration clearly has important implications for human well-being, including many mental health challenges (Yates et al., 2021; Schwerdtle et al., 2018). From a psychological standpoint, migration is not a single event with a definitive endpoint but a continual process with evolving impacts (Shultz et al., 2019). Trauma from disastrous extreme weather events could combine with anguish from the 'intolerable loss' (Handmer & Nalau, 2019) of intergenerational practices, livelihoods and connections to place (Shultz et al., 2019). Those experiences of loss can also depend upon how voluntary the migration and the ability to continue socio-cultural practices in new locations (Handmer & Nalau, 2019; Torres & Casey, 2017). More disruptive movements may give rise to chronic stress, anxiety and depression, especially if returning home is impossible (Shultz et al., 2019; Manning & Clayton, 2018; Torres & Casey, 2017; McIver et al., 2016; Britton & Howden-Chapman, 2011).

For those who stay, different stressors on ways of life and levels of resilience are at play. Food is an integral component of cultural celebrations and social cohesion, but across the Pacific climate extremes are altering the availability of particular foods used for celebrations and ceremonies (Savage et al., 2021). Furthermore, the threat of future climate extremes is described as a deterrent to invest the significant time and labour required for food growing and provides difficulties in the preparation of traditional dishes (Savage et al., 2021). These social trends have been driving the nutrition transition away from traditional, locally grown foods to a diet high in energy-dense, nutrient-poor imported foods (Savage et al., 2020; Charlton et al., 2016; James, 2016; Martyn et al., 2015). Locally grown produce, especially in urban areas, is increasingly replaced by a limited array of imported foods such as rice, bread, instant noodles and tinned fish and meat (Savage et al., 2020; James, 2016; Martyn et al., 2015). Furthermore, in vulnerable areas increased reliance on, and the expectation of, food aid has also contributed to the erosion of traditional strategies for resilience to climate extremes, such as food preservation and food stockpiling, and overall waning of food growing in areas with greater access to store-bought foods (Savage et al., 2021; Jackson, 2019; Wentworth, 2019; Campbell, 2015). Pacific climate extremes have therefore constrained the agency of people to make beneficial food choices (Savage et al., 2021). The weight of these cumulative stressors can lead to physical health issues and distress, anxiety or depression from feeling unable to support the family and wider community (Yates et al., 2021).

Conclusion

Despite strong ties to the land, Pacific Island people have a long history of migration, with the ocean acting as a highway rather than a barrier to mobility (Hau'ofa, 1993). More recently, resettlements in the Pacific suffered multiple problems and controversies and it's almost impossible to find a best practice example (Kupferberg, 2021), particularly for the present context. The years following World War II saw three significant international relocations of Pacific Island communities instigated by British colonial rule in what were deemed overcrowded islands on the Gilbert and Ellice Islands, now Kiribati and Tuvalu, respectively (Connell,

2012), all of which proved disastrous. As ex-colonial Western powers have played such a major part in past changes imposed on Pacific islands, and are responsible for the vast amount of GHGe now threatening the islands' very existence, there is a strong argument that these colonial powers now hold an obligation for future support and provision. The importance of not making the same mistakes or imposing uninformed top-down solutions is key to providing the right kind of support in the future. The discontent with newcomers in host community's observable in a wide range of Pacific resettlement schemes is something which must be addressed in any future relocation policies and projects (Kupferberg, 2021). Strengthening and nurturing regional solidarity is crucial in avoiding hostility or conflict. Meaning public opinion needs to be sought so relocations avoid the pitfall of being top-down elite projects supported by international organisations, academics and current political administrations with little base in reality and local attitudes (Kupferberg, 2021). 'Too often climate change adaptation efforts use a top-down approach and standardised models, which leaves the people facing environmental degradation with little say in the actual decision-making process' (Bertana, 2017).

Indigenous knowledge, narratives and values have traditionally been excluded in climate change scholarship (Yates et al., 2021), which is often biased towards Eurocentric research paradigms (Jones, 2019). These paradigms tend to prioritise precisely defined and empirically measured research. Such inflexibility can clash with indigenous knowledge (Alexander et al., 2011), which is 'gained through trans-generational experiences, observations, and transmission' (United Nations Environment Programme, 2021). Also, Eurocentric research tends to construct reality from a purely European and ex-colonial worldview, assuming associated values, such as anthropocentrism and individualism. These are the same values largely driving the climate crisis (Jones, 2019; Naidoo, 1996; Lala, 2015). Consequently, Eurocentrism has minimised the role of indigenous values in climate research, reducing space for Indigenous voices in decision-making (Jones, 2019) which needs to change. Encouragingly, there is growing recognition of the value of indigenous knowledge in environmental protection (Alexander et al., 2011; Etchart, 2017; Green & Raygorodetsky, 2010) and there is a push for more evidence-informed climate-related migration policies (McMichael et al., 2019; International Organisation for Migration, 2019; Wiegel et al., 2019).

Going forwards, the future of Kiribati, Tuvalu, the Republic of the Marshall Islands and other low-lying PICs is uncertain. They need to prepare for the worst, whilst preserving their dignity in the process (Kupferberg, 2021). Adaptation projects, mitigation, disaster risk reduction, temporary or permanent migration, humanitarian visas and relocation are all partial solutions to a complex problem. They cannot exist in isolation but should be viewed and approached as complimentary. As several commentators and researchers have highlighted, agreement between governments is far from enough. The involvement of local communities in concerned countries is essential.

Climate change 'adaptation in a local context requires processes to address social and cultural issues as well as climatic ones, enabling communities to deal better with environmental uncertainty in a way that suits them, without losing the value systems and practices that underpin their way of life. In short, 'adaptation' should be about adjusting to both climate and social change' (Warrick, 2012).

People's ties to their homelands shape their experiences of migration and climate change. For many Pacific cultures, land—aba (Kiribati), fenua (Tuvaluan) and āne (Marshallese) cannot be separated from culture and identity (Yates et al., 2021). Land connects past, present and future peoples. It is a marker of social standing which cannot be sold and is passed down through the generations (Hermann & Kempf, 2017; McClain et al., 2020; Thompson, 2015; Shen & Binns, 2012). There is a strong argument concerning the ethical obligation for ex-colonial and major GHGe countries to take responsibility for co-creating effective responses and actions. Traditional forms of development assistance, such as simply giving aid to poor country governments, are inadequate to the task (Rieffel, 2018; Betts & Collier, 2018).

References

Aburn, A. and Wesselbaum, D. (2017). 'Gone with the Wind: International Migration', University of Otago Economics Discussion Papers No. 1708, Vol. 25.

Ahmed, S. (2009). 'From Underwater, Maldives Sends Warning on Climate Change', CNN, October 17, 2009, [online]. Available at: http://edition.cnn.com/2009/WORLD/asiapcf/10/17/maldives.underwater.meeting/ (Accessed September 03, 2021)

Alexander, C. Bynum, N. Johnson, E. King, U. Mustonen, T. Neofotis, P. Oettlé, N. Rosenzweig, C. Sakakibara, C. Shadrin, V. Vicarelli, M. Waterhouse, J. and Weeks, B. (2011). Linking Indigenous and scientific knowledge of climate change. Bioscience, 61(6), pp. 477–484.

Anderson, C. (2017). 'New Zealand Considers Creating Climate Change Refugee Visas', *The Guardian*, October 31, 2017, [online]. Available at: https://www.theguardian.com/world/2017/oct/31/new-zealand-consid ers-creating-climate-change-refugee-visas (Accessed September 01, 2021)

Bamber, J. L. et al. (2019). 'Ice Sheet Contributions to Future Sea-level Rise from Structured Expert Judgment', PNAS 116, no. 23.

Beeler, C. (2018). UN compact recognizes climate change as driver of migration for first time (radio program). The World. PRX/WGBH. [online]. Available at: https://www.pri.org/stories/2018-12-11/un-compact-recognizes-cli mate-change-driver-migration-first-time (Accessed September 03, 2021)

Bertana, A. (2017). 'How a Community in Fiji Relocated to Adapt to Climate Change', Scholars Strategy Network, [online]. Available at: https://sch olars.org/contribution/how-community-fiji-relocated-adapt-climate-change (Accessed September 02, 2021)

Betts, A. and Collier, B. (2018). 'How Europe Can Reform Its Migration Policy', *The Importance of Being Sustainable*. [online]. Available at: https://www.foreignaffairs.com/articles/europe/2018-10-05/how-europe-can-reform-its-migration-policy (Accessed September 01, 2021)

Britton, E. and Howden-Chapman, P. (2011). 'The Effect of Climate Change on Children Living on Pacific islands'. *Clim Chang Rural Child Heal*, 26(1), pp. 177–188.

Burch, E. (2020): 'A Sea Change for Climate Refugees in the South Pacific: How Social Media – Not Journalism – Tells Their Real Story', Environmental Communication.

Campbell, J. and Warrick, O. (2014). 'Climate Change and Migration Issues in the Pacific', *Disaster Displacement*, Vol. 2, [online]. Available at: https://disasterdisplacement.org/wp-content/uploads/2015/03/2.-Climate-Change-and-Migration-Issues-in-the-Pacific.pdf (Accessed September 04, 2021)

Campbell, J. (2015). 'Development, Global Change and Traditional Food Security in Pacific Island countries'. *Regional Environmental Change* 15, pp. 1313–1324.

Campbell in J. McAdam, ed., (2010). *Climate Change and Displacement: Multidisciplinary Perspectives*. Hart Publishing, pp. 59–62.

Charlton, K.E. Russell, J. Gorman, E. Hanich, Q. Delisle, A. Campbell, B. and Bell, J. (2016). 'Fish, Food Security and Health in Pacific Island Countries and Territories: A Systematic Literature Review'. *BMC Public Health* 16, pp. 285.

Connell, J. (2007). 'Population Resettlement', 130; and Pireport, 'Relocated Fiji Villagers Want Rabi Island Back', [online]. Available at: http://www.pireport.org/articles/2007/06/05/relocated-fiji-villagers-want-rambi-island-back (Accessed August 18, 2021)

Connell, J. (2012). 'Population Resettlement in the Pacific: Lessons from a Hazardous History?' *Australian Geographer* 43, no. 2 pp. 127.

COP23, (2017). 'Kiribati' Current forecast: Kiribati and a changing climate [online]. Available at: https://cop23.com.fj/kiribati/ (Accessed August 25, 2021)

Curtain, R. et al. (2016). 'Pacific Possible: Labour Mobility: The Ten Billion Dollar Prize', The World Bank, Vol. i–ii, [online]. Available at: http://pubdocs.worldbank.org/en/555421468204932199/pdf/labour-mobility-pacific-possible.pdf (Accessed August 30, 2021)

Dempster, H. and Ober, K. (2020). 'New Zealand's "Climate Refugee" Visas: Lessons for the Rest of the World', Devpolicy, [online]. Available at: https://devpolicy.org/new-zealands-climate-refugee-visas-lessons-for-the-rest-of-the-world-20200131/ (Accessed August 31, 2021)

Drinkall, S. Leung, J. Bruch, C. Micky, K. and Wells, S. (2019). 'Migration with dignity: A case study on the livelihood transition of micronesians to Portland and Salem, Oregon'. *Journal of Disaster Research*, 14(9), pp. 1267–1276.

Ebi, K.L. and Bowen, K. (2016). 'Extreme Events as Sources of Health Vulnerability: Drought as an Example'. *Weather Climate Extremes* Vol. 11, pp. 95–102.

Emont, J. and Anandarajah, G. (2017). 'Rising waters and a smaller island: What should physicians do for Tuvalu'? *Ama Journal of Ethics*, 18(12), pp. 1211–1221.

Etchart, L. (2017). 'The role of indigenous peoples in combating climate change'. *Palgrave Communications*, 3(1), pp. 1–4.

Farbotko, C. and Lazarus, H. (2012). 'The first climate refugees? Contesting global narratives of climate change in Tuvalu'. *Global Environmental Change*, Vol. 22(2), pp. 382–390.

Ferris, E. Cernea, M.M. and Petz, D. (2011). 'On the Front Line of Climate Change and displacement: Learning from and with Pacific Island Countries', *The Brookings Institution – London School of Economics Project on Internal Displacement*, Vol 25.

Quoted by Gemenne (2009) in C.E. Kenfack, 'Climate Migrants: Victims and Actors of Environmental Violence', *Human Welfare* 4 (2015). Vol. 12. [online]. Available at: https://climig.com/climate-migrants-victims-and-actors-of-environmental-violence/ (Accessed August 31, 2021)

Gemenne, F. (2010). 'Tuvalu, un laboratoire du changement climatique? Une critique empirique de la rhétorique des "canaris dans la mine'. *Revue Tiers Monde*, Vol. 204, pp. 89–107.

Gillard, M. and Dyson, L. (2012). 'Kiribati migration to New Zealand: Experience, needs and aspirations'. *Impact Research*.

The Government Office for Science, (2011). 'Foresight: Migration and Global Environmental Change – Final Project Report', Government of the UK, Vol. 9. [online]. Available at: https://www.gov.uk/government/publications/migration-and-global-environmental-change-future-challenges-and-opportunities (Accessed September 02, 2021)

Green, D. and Raygorodetsky, G. (2010). 'Indigenous Knowledge of a Changing Climate'. *Climatic Change*, Vol. 100(2), pp. 239–242.

Handmer, J. and Nalau, J. (2019). Loss and Damage From Climate Change. Chapter 15, Understanding Loss and Damage in Pacific Small Island Developing States; pp. 365–81.

Hau'ofa, E. (1993). 'Our Sea of Islands'; in In a New Oceania: Rediscovering Our Sea of Islands, ed. E. Waddell, Suva: University of the South Pacific.

Hermann, E. and Kempf, W. (2017). 'Climate Change and the Imagining of Migration: Emerging Discourses on Kiribati's land purchase in Fiji'. *The Contemporary Pacific*, Vol. 29(2), pp. 231–263.

Initiative, T. N. (2015). 'Cross-Border Displaced Persons in the Context of Disasters and Climate Change'. Vol. 1. [online]. Available at: https://nanseninitiative.org/wp-content/uploads/2015/02/PROTECTION-AGENDA-VOLUME-1.pdf (Accessed August 31, 2021)

International Organisation for Migration (IOM). (2019). 'Standing Committee on Programme and Finance: Twenty-Fourth Session - Update on Policies and Practices Related to Migration, the Environment and Cli- mate Change and IOM's Environmental Sustainability Programme'. [online]. Available at: https://governingbodies.iom.int/system/files/en/scpf/24th/S-24-5-Update%20on%20policies%20and%20practices%20related%20to%20MECC_0.pdf (Accessed August 31, 2021)

Ionesco, D. (2019). 'Let's talk about climate migrants, not climate refugees. UN Sustainable Development'. [online]. Available at: https://www.un.org/sustainabledevelopment/blog/2019/06/lets-talk-about-climate-migrants-not-climate-refugees/ (Accessed August 30, 2021)

IPCC, (2001). 'Climate Change: Impacts, Adaption, and Vulnerability', Vol. 867. [online]. Available at: https://www.ipcc.ch/report/ar3/wg2/ (Accessed August 30, 2021)

IPCC, (2014). 'Climate Change 2014: Synthesis Report', Vol. 4. [online]. Available at: https://www.ipcc.ch/site/assets/uploads/2018/05/SYR_AR5_FINAL_full_wcover.pdf (Accessed August 30, 2021)

IPCC. (2019a). Ar6 Synthesis Report: Climate change 2022. [online]. Available at: https://www.ipcc.ch/report/sixth-assessment-report-cycle/ (Accessed August 30, 2021)

IPCC, (2019b). 'Summary for Policymakers'. 'In: *IPCC Special Report on the Ocean and Cryosphere in a Changing Climate*', Vol. 20. [online]. Available at: https://www.ipcc.ch/site/assets/uploads/sites/3/2019/11/03_SROCC_SPM_FINAL.pdf (Accessed August 31, 2021)

Jackson, G. (2019). 'The Influence of Emergency Food Aid on the Causal Disaster vulnerability of Indigenous Food Systems'. *Agriculture and Human Values*.

James, S. (2016). 'Urban Nutrition Program for El Nin¯o Drought Response Project: Final Report'. Wan Smol Bag Theatre & World Vision (Port Vila, Vanuatu).

Jones, R. (2019). 'Climate Change and Indigenous Health Pro- Motion'. *Global Health Promotion*, Vol. 26(3), pp. 73–81.

Kiribati Office of Climate Change, (2009). 'Kiribati Reveals "Human Face of Climate Change"' Press release, [online]. Available at: http://www.climate.gov.ki/2009/12/14/kiribati-reveals-the-human-face-of-climate-change-to-cop16/ (Accessed August 30, 2021)

Kupferberg, J. S. (2021). 'Migration and Dignity – Relocation and Adaptation in the Face of Climate Change Displacement in the Pacific – A human rights perspective', *The International Journal of Human Rights*.

Krajick, K. (2018). 'Climate Migrants Will Soon Shift Populations of Many Countries, Says World Bank', Earth Institute – Columbia University. [online]. Available at: https://blogs.ei.columbia.edu/2018/03/19/climate-refugees-will/ (Accessed August 30, 2021)

Lala, J. M. (2015). 'A Phenomenological Exploration of the Psychological Impacts of Climate Change; A Focus on Funafuti', Tuvalu [doctoral dissertation]. University of the South Pacific.

Locke, J. T. (2009). 'Climate Change-Induced Migration in the Pacific region: Sudden Crisis and Long-Term Developments'. *Geographical Journal*, Vol. 175(3), pp. 171–180.

Maekawa, M. Singh, P. Charan, D. Yoshioka, N. and Uakeia, T. (2019). Livelihood re-establishment of emigrants from Kiribati in Fiji. Journal of Disaster Research, Vol. 14(9), pp. 1277–1286.

Malua, S. (2014). *The Tuvalu community in Auckland: A focus on health and migration*. Auckland: School of Social Sciences, the University of Auckland. Transnational Pacific Health Through the Lens of Tuberculosis Report No.: 4. [online]. Available at: https://www.auckland.ac.nz/en.html (Accessed August 31, 2021)

Manch, T. (2018). 'Humanitarian visa Proposed for Climate Change Refugees Dead in the Water', Stuff. [online]. Available at: https://www.stuff.co.nz/environment/106660148/humanitarian-visa-proposed-for-climate-change-refugees-dead-in-the-water (Accessed August 30, 2021)

Manning, C. and Clayton, S. (2018). *Psychology and climate change*. Elsevier. pp. 217–244.

Martyn, T. Yi, D. and Fiti, L. (2015). *Identifying the household factors, and food items, most important to nutrition in Vanuatu*. Food and Agriculture Organization of the United Nations, Vanuatu.

McAdam, J. (2012). *Climate Change, Forced Migration, and International Law*. Oxford; New York: Oxford University Press, Vol. 105.

McClain, S. N. Bruch, C. Nakayama, M. and Laelan, M. (2020). 'Migration with dignity: A case study on the Livelihood Transition of Marshallese to Springdale, Arkansas'. *Journal of International Migration and Integration*, Vol. 21, pp. 847–859.

McClain, S. N. Seru, J. and Lajar, H. (2019). 'Migration, transition, and livelihoods: A comparative analysis of Marshallese pre- and post- Migration to the United States'. *Journal of Disaster Research*, Vol. 14(9), pp. 1262–1266.

McIver, L. Kim, R. Woodward, A. Hales, S. Spickett, J. and Katscherian, D. (2016). 'Health Impacts of Climate Change in Pacific island countries: A regional assessment'. *Environmental Health Perspectives*, Vol. 124(11), pp. 1707–1714.

McMichael, C. Katonivualiku. M. and Powell, T. (2019). 'Planned Relocation and Everyday Agency in Low-lying Coastal Villages in Fiji'. *The Geographical Journal*, Vol. 185(3), pp. 325–337.

Miller, S.D. (2018). 'Xenophobia Toward Refugees and Other Forced Migrants', World Refugee Council Research Paper no. 5, Vol. 1. [online]. Available at: https://reliefweb.int/sites/reliefweb.int/files/resources/WRC%20Research%20Paper%20no.5.pdf (Accessed August 30, 2021)

Naidoo, A. (1996). 'Challenging the hegemony of Eurocentric psychology'. *Journal Community Heal Science*, Vol. 2(2), pp. 9–16.

Oxfam. (2019). Forced from home: Climate-fuelled displacement. Oxfam media briefing. [online]. Available at: https://oxfamilibrary.openrepository.com/bitstream/handle/10546/620914/mb-climate-displacement-cop25-021219-en.pdf (Accessed August 30, 2021)

Pashley, A. (2016). 'Kiribati President: Climate-induced Migration is 5 Years Away', Climate Change News, [online]. Available at: https://www.climatechangenews.com/2016/02/18/kiribati-president-climate-induced-migration-is-5-years-away/ (Accessed August 30, 2021)

Purvis, K. (2016). 'Sinking States: The Islands Facing the Effects of Climate Change', The Guardian, [online]. Available at: https://www.theguardian.com/global-development-professionals-network/gallery/2016/feb/15/pacific-islands-sinking-states-climate-change (Accessed August 30, 2021)

Ray, C. (2019). 'Rejecting Reality: Kiribati's Shifting Climate Change Policies', The University of Texas. [online]. Available at: https://sites.utexas.edu/climatesecurity/2019/12/31/kiribati-policy-shift/ (Accessed August 30, 2021)

Rieffel, L. (2018). 'The Global Compact on Migration: Dead on arrival?' Brookings. [online]. Available at: https://www.brookings.edu/blog/up-front/2018/12/12/the-global-compact-on-migration-dead-on-arrival/ (Accessed August 31, 2021)

Robie, D. (2020). From Nuclear Refugees to Climate Justice—The Rainbow Warrior legacy. Auckland University of Technology Pacific Media Centre: Asia Pacific Report. [online]. Available at: https://asiapacificreport.nz/2020/07/10/from-nuclear-refugees-to-climate-justice-the-rainbow-warrior-legacy/ (Accessed August 31, 2021)

Roman, M. T. (2013). Migration, Transnationality, and Climate Change in the Republic of Kiribati [doctoral dissertation on the Internet]. University of Pittsburgh. [online]. Available at: https://core.ac.uk/download/pdf/19522843.pdf (Accessed August 31, 2021)

Savage, A. Bambrick, H. and Gallegos, D. (2020). 'From Garden to Store: Local Perspectives of ChangingFood and Nutrition Security in a Pacific Island Country'. Food Security.

Savage, A. Bambrick, H. and Gallegos, D. (2021). 'Climate Extremes Constrain Agency and Long-Term Health: A Qualitative Case Study in a Pacific Small Island Developing State'. *Weather and Climate Extremes*, Vol. 31.

Schwerdtle, P. Bowen, K. and McMichael, C. (2018). 'The Health Impacts of Climate-Related Migration'. *BMC Medicine*, Vol. 16(1), pp. 1–7.

Shen, S. and Binns, T. (2012). 'Pathways, Motivations and Challenges: Contemporary Tuvaluan Migration to New Zealand'. *GeoJournal*, Vol. 77 (1), pp. 63–82.

Shen, S. and Gemenne, F. (2011). 'Contrasted views on environmental change and migration: The case of Tuvaluan migration to New Zealand'. *International Migration (geneva, Switzerland)*, Vol. 49(s1), pp. e224– e242.https://doi.org/10.1111/j.1468-2435.2010.00635.x

Shultz, J. M. Rechkemmer, A. Rai, A. and McManus, K. T. (2019). 'Public health and mental health implications of environmentally induced forced migration'. *Disaster Medicine and Public Health Preparedness*, Vol. 13(2), pp. 116–122.

Sinha, T. (2020) 'What are the Cause of Poverty in Kiribati?' The Borgen Project, n.d., [online]. Available at: https://borgenproject.org/causes-of-poverty-in-kiribati/ (Accessed August 31, 2021)

Siose, L. (2017). Community perception on migration as an adaptation strategy to the impact of climate change in Tuvalu: the case of communities in Tuvalu and migrated communities in New Zealand [dissertation on the Internet]. The University of the South Pacific. [online]. Available at: http://digilib.library.usp.ac.fj/cgi-bin/library.cgi?e=d-01000-00---off-0usplibr1--00-1----0-10-0---0---0direct-10---4-------0-1l--11-en-50---20-about---00-3-1-00-0--4--0--0-0-11-10-0utfZz-8-00&a=d&c=usplibr1&cl=CL2.10.18 (Accessed August 31, 2021)

SPREP, 'Factsheet – Pacific Climate Change', SPREP, 2015, [online]. Available at: https://www.sprep.org/attachments/Publications/FactSheet/pacificclimate.pdf (Accessed August 31, 2021)

Su, Y. (2020). UN Ruling on Climate Refugees Could be Gamechanger for Climate Action. Climate Change News. [online]. Available at: https://www.climatechangenews.com/2020/01/29/un-ruling-climate-refugees-gamechanger-climate-action/ (Accessed August 31, 2021)

Thompson, M. A. (2015). The Settlement Experiences of Kiribati migrants living in New Zealand (doctoral dissertaton on the Internet). [online]. Available at: University of Otago. https://ourarchive.otago.ac.nz/handle/10523/6144 (Accessed August 31, 2021)

Torres, J. M. and Casey, J. A. (2017). The Centrality of Social Ties to Climate Migration and Mental Health. *BMC Public Health*, Vol. 17(1), pp. 1–10.

UN, (2019). 'World Population Prospects 2019'. [Available online] https://population.un.org/wpp/Publications/Files/WPP2019_DataBooklet.pdf (Accessed August 31, 2021)

United Nations Environment Programme. (2021). 'Indigenous peoples and their communities'. [online]. Available at: https://www.unep.org/pt-br/node/21757 (Accessed August 31, 2021)

UNHCR. (2020). 'Climate change and disaster displacement'. The UN High Commission on Refugees. [online]. Available at: www.unhcr.org/en-us/climate-change-and-disasters.html (Accessed August 31, 2021)

Walker, B. (2017). 'An Island Nation Turns Away from Climate Migration, Despite Rising Seas', Inside Climate News, November 20, 2017. [online]. Available at: https://insideclimatenews.org/news/20112017/kiribati-climate-change-refugees-migration-pacific-islands-sea-level-rise-coconuts-tourism (Accessed August 31, 2021)

Walters, L. (2019). 'New Plan Gives Pacific People Chance to Stay Home', Newsroom, April 25, 2019. [online]. Available at: https://www.newsroom.co.nz/groundbreaking-project-gives-pacific-people-chance-to-stay-in-their-homeland (Accessed August 31, 2021)

Warrick, O. (2012). *Climate change and social change: vulnerability and adaptation in rural Vanuatu*. The University of Waikato, New Zealand. [online]. Available at: http://webistem.com/psi2009/output_directory/cd1/Data/articles/000141.pdf (Accessed August 31, 2021)

Watkin, S. Foon, M. and Liddell, S. (2019). 'Temaiku Land and Urban Development – Building Sustainable Climate Change Resilience for Kiribati' (Australasian Coasts and Ports 2019 Conference: Future directions from 40 [degrees] S and beyond, Hobart, 10–13 September 2019, Engineers Australia, 2019), pp. 1204–10,

Webb, J. (2020). 'Kiribati Economic Survey: Oceans of Opportunity', Asia & the Pacific Policy Studies 7 Vol. 5. [online]. Available at: https://onlinelib rary.wiley.com/doi/epdf/10.1111/eci.13484 (Accessed August 31, 2021)

Wentworth, C. (2019). 'Unhealthy aid: food security programming and disaster responses to cyclone Pam in Vanuatu'. Anthropology Forum.

Wiegel, H. Boas, I. and Warner, J. (2019). 'A mobilities perspective on migration in the context of environmental change'. *Wiley Interdisciplinary Reviews: Climate Change*, Vol. 10(6), pp. 1–9.

Wodon, Q. Liverani, A. and Joseph, G. (2014). 'Climate Change and Migration: Evidence from the Middle East and North Africa', World Bank Publications, 2014, xiii. [online]. Available at: https://ebookcentral.proquest.com.pro quest.com/lib/malmo/detail.action?docID=1757562 (Accessed August 31, 2021)

World Bank, (2019). 'Kiribati', The World Bank. [online]. Available at: https://data.worldbank.org/country/KI (Accessed August 31, 2021)

Yates, O.E.T. Manuela, S. Neef, A. and Groot, S. (2021). 'Reshaping ties to land: a systematic review of the psychosocial and cultural impacts of Pacific climate-related mobility', *Climate and Development*.

Sustainable Development from Unsustainable Climate: Sustainable Development Goals and the Pacific Small Island Developing States

Sojin Lim◉

Climate change and environmental issues cannot be emphasised enough in the twenty-first century. At almost every turn, the international agenda embraces issues of climate change and related natural disasters. For instance, we have experienced a greater number of extreme heatwaves and floods than ever before across the globe. Since the 1990s, the international community has established a consolidated effort to address ever-increasing climate consequences. The United Nations Conference on Environment and Development (UNCED), also known as the Earth Summit, was held in Rio de Janeiro, Brazil, in 1992 on the global environment, and member countries agreed to establish the United Nations Framework Convention on Climate Change (UNFCCC) at this conference.

When the third world environment conference—called Rio+20—was held, again in Rio de Janeiro, in 2012, the government of Colombia

S. Lim (✉)
University of Central Lancashire, Preston, UK
e-mail: SLim4@uclan.ac.uk

presented the 'Rio+20 Sustainable Development Goals (Rio+20 SDGs)' zero draft. This was the very first attempt by a developing country to suggest development goals that reflected the challenges facing developing countries at the international level. As the zero draft did not include specific targets or timelines, the UN created an open working group (OWG) to develop the draft further. The OWG presented 17 goals and 169 indicators, drawn up over the course of eight meetings, from the Colombian zero draft. Discussion on these Rio+20 goals and indicators, at first, took place separately from the post-2015 development framework discussion on creating a new set of global goals to succeed the Millennium Development Goals (MDGs). Eventually, though, the 17 Rio+20 goals and 169 indicators developed by the OWG were incorporated into the post-2015 development framework, as stakeholders realised that the issues surrounding climate change and the environment could not be discussed apart from others such as economic and social development. Also, there were quite a few overlapping areas between the Rio+20 SDGs and the post-2015 development framework discussion. In light of this, UN members adapted and incorporated the 17 SDGs and 169 indicators into the 'UN Sustainable Development Agenda' in 2015 (Lim 2016). Even though the SDGs originated from the environment and climate change-focused conference, they have now been expanded to address the five main principles of people, planet, prosperity, peace and partnership within the sustainable development agenda's framework (see UN 2015).

As such, the UN Sustainable Development Agenda, adopted in 2015 by all member states, clearly embeds the Rio principles in its Article 11 by stating: "We reaffirm the outcomes of all major UN conferences and summits which have laid a solid foundation for sustainable development and have helped to shape the new agenda. These include the Rio Declaration on Environment and Development, the World Summit on Sustainable Development, ... and the UN Conference on Sustainable Development" (UN 2015: 4). In Article 11, UN members also reassured the follow-ups to these conferences, including the third International Conference on Small Island Developing States (SIDS) (UN 2015: 4–5). Further, when the UN Sustainable Development Agenda of 17 SDGs was presented, it indicated that its Goal 13 on climate change and its impact was in line with the UNFCCC as 'the UNFCCC is the primary international, intergovernmental forum for negotiating the global response to climate change' (UN 2015: 14). It also recognised the SIDS Accelerated Modalities of Action (SAMOA) Pathway in its Article 64 (UN 2015: 28).

With that in mind, this chapter aims to examine how SIDS climate change challenges have been addressed in UN SDG implementation policies in the case of the Pacific region. While there is an extensive amount of research about SIDS and climate change to be found in the academic literature, few have given particular attention to the SDGs and donor engagement in SIDS. Most existing research discusses what has happened in SIDS due to climate change, how unjustifiable or unequal this impact has been on people living in the small islands vis-à-vis the world, how SIDS have adapted so far and what their strategies should be in the face of future challenges, which are more specifically analysed in the following section. However, this body of work has not fully explored how international financial and technical support has been provided, and should be provided in the future, in order to preserve these places and enhance resilience. Accordingly, this chapter navigates how Pacific SIDS have reflected unsustainable situations, caused by climate change, into their national development strategies, and thus, SDG implementation policies, and further asks whether the donor community integrates the needs raised by the recipients into their development aid policies. With regard to the latter, the intention is to ascertain whether the North's rhetoric on and perceptions of SIDS and climate change neglect, or adhere to, actual situations portrayed in islanders' narratives.

In doing so, this chapter reflects Walshe and Stancioff's (2018) argument that discussion on SIDS tends to contribute to the tendency towards simplifying the diversity of small islands and their complex problems with climate change. In other words, this chapter does not attempt to generalise the situations that SIDS in the Pacific region face, but rather to analyse each country-case in the context of the SDGs. This does not mean that the chapter denies the widely agreed common finding from existing research that SIDS are pure victims of climate change, with the least contribution to the world's carbon emissions, for which most of the North is responsible (see, for example, Halsted 2016; Kelman 2010; Rudiak-Gould 2014; Walshe and Stancioff 2018). But rather it intends to explore individual Pacific island's development policies in the context of the SDGs, and how donor approaches are commensurate with islanders' priorities.

The chapter is organised in six sections. Following this introduction (Section 1), Section 2 reviews existing literature on SIDS and climate change, in order to identify the conventional belief or emphasis in the existing academic discussion vis-à-vis the related themes. Section 3 explores the UN SDGs and assesses how SIDS climate change and development issues have been reflected in the SDGs. In Section 4, the chapter analyses the SDGs Voluntary National Review (VNR) of each SIDS, through which we can learn whether the Pacific SIDS development agenda, including climate change mitigation and adaptation, has been embedded in the UN SDGs. Also, this analysis demonstrates the Pacific islands' development policies and narratives of climate change influence. Next, Section 5 examines how the donor community addresses the Pacific SIDS climate change and sustainable development agenda in implementing the UN SDGs, based on the case of Australia. By so doing, this chapter is able to assess whether donors consider the UN SDGs and SIDS commitment, as well as recipient countries' policies. Also, the chapter looks at whether theories of SIDS and climate change found in the existing literature are reflected in actual donor policy.

Official development assistance (ODA) or development aid is categorised between multilateral aid and bilateral aid, and is composed of grant and concessional aid. According to the Organisation for Economic Co-operation and Development (OECD)—traditional donors are members of its Development Assistance Committee (DAC)—donors have provided more grants than loans to SIDS, and more than half of the amount has been given by bilateral donors than by multilateral organisations (OECD 2021). The top five donors to SIDS are Australia, the United States (US), the European Union (EU), France and the International Development Association (IDA) (OECD 2021). In the Pacific region, the dominant single donor is Australia in Melanesia, and Australia and New Zealand in the Polynesian SIDS, while the US is the largest single donor in the northern Pacific compact territory. Among these three countries, Australia shows the largest ODA contribution across the Pacific (UNDP 2017). However, the statistics do not include non-DAC member donors' financial support, as they are not required to provide ODA data to the OECD or other relevant international organisations. Accordingly, this chapter focuses on the case of Australia as a donor, given that it has more influence in the region than other donors.

Based on this, the final section (Section 6) of the chapter discusses linkages or disparities between theory (SIDS climate change discussion and SDGs commitment to SIDS) and practice (Pacific islands' sustainable development policies and Australian donor policy). That is, the main focus of this chapter is on examining how climate change in the Pacific region has been interpreted in academic discussions, the global norm of sustainable development, donor aid policy and the Pacific SIDS sustainable development agenda. In conclusion, this chapter argues that SIDS in the Pacific have faced specific vulnerabilities, which have made them lag behind in international development cooperation, even though the SDGs platform promotes the value of 'leave no one behind'. The chapter also challenges the conventional argument that Pacific SIDS prioritise climate change and related natural disasters, and argues instead that these small countries are not very different from other developing countries in the sense that economic and social development are at the core of their sustainable development efforts, even as they carry a greater burden for resilience capacity development, given that they are directly experiencing ocean-related climate change consequences.

SMALL ISLAND DEVELOPING STATES AND CLIMATE CHANGE

About 58 countries in three regions—the Caribbean, the Pacific, and the Atlantic, Indian Ocean, Mediterranean and South China Seas (AIMS)—have been recognised as most at-risk countries from climate change and were defined as SIDS during the 1992 UNCED (Thomas et al. 2020). Since the concept of SIDS was created in 1992, these 58 countries have been recognised as having common problems that owe to climate change—for example, land loss and, as a result, food insecurity due to sea-level rise. Not only territorial reduction but also economic downturn has led to forced migration and displacement and thus loss of heritage, indigenous culture, and further sovereignty. Also, negative impacts on tourism and mental health have been related issues in these countries. Overall, SIDS climate change resilience is at risk.

A majority of the existing literature portrays rising sea levels and vanishing islands as the main issues caused by climate change. One of the most common themes related to sinking islands is migration and displacement due to sea-level rise. Most people in SIDS live and work near coastal regions. Thus, these people face losing their homes when small islands become even smaller. For example, people in countries like Tonga, Kiribati, the Maldives, the Marshall Islands, the Federated States of Micronesia (FSM), and Tuvalu have already lost large parts of their habitable land (Kelman 2010). It is not difficult to reason further that limited land availability in small islands affects physical industrial land capacity, and ultimately the economy itself.

Here, scholars, including Halsted (2016), argue that people displaced from SIDS due to climate change should be defined as 'environmental refugees' and seen as global citizens, and thus requiring of international support. However, academics like Perumal (2018: 46) contest that 'climate refugee' is a rather 'sensationalised and over-simplistic' term and argue that using the term 'climate migrant' is more appropriate. Contrary to conventional belief, in some cases, local people, for instance in Kiribati, reject becoming climate refugees, but rather try to determine their own terms by requesting external assistance so that they can 'migrate' (Kelman 2018). This chapter does not intend to contribute to this debate, but uses the expressions 'climate migration' or 'displacement' as common ground.

Fear of the possible future impacts of climate change, or experience of cultural change and place-based identity conflict, have affected islanders and created a need for prioritisation of their mental health and well-being in the preventive and treatment-related health system (Gibson et al. 2020; Kelman et al. 2021). In the case of Kiribati, climate migration-related anxiety of being landless and uncertain recovery from natural disasters has left scars in people's minds, along with conflicts with local residents where they have been relocated (Donner 2015). People living in SIDS have also faced physical health risks. For example, lack of fresh local food availability has meant that islanders have had to compromise their food consumption by greater dependence on imported foods that are mostly less healthy and eventually cause diet-related non-communicable diseases (NCDs), such as obesity and diabetes (Mclver et al. 2016; Savage et al. 2020). Extreme weather events and gradual environmental change, as well, have affected both the mental and physical health of people in SIDS (Gibson et al. 2020; Mclver et al. 2016).

Food and nutrition insecurities, which are the main reasons for changes in islanders' diets, have been another secondary effect associated with climate change in SIDS. Extreme weather events and sea-level rise have challenged agriculture and fishery as livelihoods, and thus the resilience and sustainability of food systems in SIDS (Lowitt et al. 2015). For example, in the Solomon Islands, fruiting patterns have changed, and marine life has been observed moving southward and eastward. As mentioned above, lack of access to local food has brought an inevitable reliance on food imports, which has been one of the main causes of local diet-related NCDs. Also, saltwater intrusion has threatened water security, including farming systems and food production that are associated with food security in turn (Gohar et al. 2019).

As seen, there are some common themes when we discuss SIDS and climate change; however, this does not mean that SIDS have only to endure the negative aspects of climate change. Coral reefs are not endangered in all cases, and can be resilient to sea-level rise and ecosystem changes. Dying corals are not necessarily the result of climate change, but can have been caused by local mining, as in the cases of the Maldives and Kiribati, or by tourism, as in the case of Barbados (Kelman 2018). When 'smallness' was defined back in 1992, it was an important feature of SIDS. About 30 years on, the demand for addressing the different vulnerabilities of each nation is greater. Some SIDS have higher gross domestic products (GDPs) than other economies (Walshe and Stancioff 2018). The small size of a population itself is not a disadvantage, but rather it makes people sensitive to climate change and its consequences faster and in a larger way (Kelman 2010). The iconic image of the 'vulnerability' of these small islanders against climate change does not always mean lack of resilience. SIDS characteristics can bring resilience to climate change, as local knowledge and their own indigenous experience can correct existing misplaced policies and practices. For example, Savage et al. (2020) imply that in overcoming food and nutrition insecurity, SIDS governments can avoid imported foods by encouraging climate-resilient cultivars and increasing accessibility to pelagic fish.

At the same time, there can be diverse urgent issues. For instance, people living in Tuvalu, Samoa, and Tonga have perceived that drought, cyclones, and other water-related issues are the most concerning problems in their communities, as shown by a survey conducted among local residents in the Pacific islands (Walshe and Stancioff 2018). As a matter of fact, sea-level change is not the only major driver of migration; extreme natural disasters, which have become more frequent in SIDS due to climate change, have also affected island economies dramatically (Halsted 2016). However, due to their smallness, addressing the issues facing each SIDS individually is not necessarily incorporated into wider policy and distant decision-making processes, such as at international conferences or in UN discussions, and sometimes not even at the level of SIDS governments. Thus, the more recent trend is towards reflecting local voices in policymaking processes (Kelman 2010), and this is also in line with the SDGs core value of 'leave no one behind'. The following section discusses this in more detail by looking at how local SIDS situations have been incorporated into the global sustainable development agenda.

SIDS Climate Change and the UN Sustainable Development Agenda

The international community has emphasised that climate change is a serious challenge and thus an obstacle for countries in achieving sustainable development. When it comes to SIDS, according to UNCED, 'SIDS, and islands supporting small communities are special case both for environment and development. They are ecologically fragile and vulnerable. Their small size, limited resources, geographic dispersion and isolation from markets, place them at a disadvantage economically and prevent economies of scale' (Scandurra et al. 2018: 382). The sustainable development agenda agreed by UN members in 2015 also clearly reflects the situation SIDS face as Article 14 states: 'increases in global temperature, sea level rise, ocean acidification and other climate change impacts are seriously affecting coastal areas and low-lying coastal countries', including SIDS (UN 2015: 5). UN members are aware of the existence of inequalities among countries, and the fact that countries like SIDS, for instance, were left behind in meeting the MDGs (UN 2015: 5). They have also recognised differentiated challenges at each country, and defined SIDS as one of the most vulnerable countries (UN 2015: 7 and 13). Table 1 shows all SDGs 17 goals, along with specific indicators that deal with SIDS.

As seen in Table 1, UN members have not recognised differentiated challenges for all 17 SDGs, but have addressed challenges related to SIDS in eight goals: health, education, energy, infrastructure, inequality, climate change, life below water and global partnership (financial support for developing countries). Also, even though the SDGs do not specifically address issues related to economic growth and employment in SIDS, Article 68 of the UN Sustainable Development Agenda urges World Trade Organisation (WTO) member countries to make consolidated efforts towards trade-related capacity-building in countries like SIDS (UN 2015: 29).

UN SDGs and the Pacific Small Islanders

Not all SIDS in the Pacific are fragile at the gross national income (GNI) level. Rather, they are mostly either middle-income countries (MICs) or even high-income countries (HICs). As seen in Fig. 1, only one country (Tonga) is a low-income country (LIC), while Nauru and Palau are HICs.

Table 1 SIDS in the UN SDGs

SDG goals	SIDS-specific indicators
Goal 1 End poverty in all its forms everywhere	—
Goal 2 End hunger, achieve food security and improved nutrition and promote sustainable agriculture	—
Goal 3 Ensure healthy lives and promote well-being for all at all ages	3.c Substantially increase health financing and the recruitment, development, training and retention of the health workforce in developing countries, especially in least developed countries and SIDS
Goal 4 Ensure inclusive and equitable quality education and promote lifelong learning opportunities for all	4.b By 2020, substantially expand globally the number of scholarships available to developing countries, in particular least developed countries, SIDS and African countries, for enrolment in higher education, including vocational training and information and communications technology, technical, engineering and scientific programmes, in developed countries and other developing countries 4.c By 2030, substantially increase the supply of qualified teachers, including through international cooperation for teacher training in developing countries, especially least developed countries and SIDS
Goal 5 Achieve gender equality and empower all women and girls	—
Goal 6 Ensure availability and sustainable management of water and sanitation for all	—

(continued)

Table 1 (continued)

SDG goals	SIDS-specific indicators
Goal 7 Ensure access to affordable, reliable, sustainable and modern energy for all	7.b By 2030, expand infrastructure and upgrade technology for supplying modern and sustainable energy services or all in developing countries, in particular least developed countries, SIDS and landlocked developing countries, in accordance with their respective programmes of support
Goal 8 Promote sustained, inclusive and sustainable economic growth, full and productive employment and decent work for all	–
Goal 9 Build resilient infrastructure, promote inclusive and sustainable industrialisation and foster innovation	9.a Facilitate sustainable and resilient infrastructure development in developing countries through enhanced financial, technological and technical support to African countries, east developed countries, landlocked developing countries and SIDS
Goal 10 Reduce inequality within and among countries	10.b Encourage ODA and financial flows, including foreign direct investment, to states where the need is greatest, in particular least developed countries, African countries, SIDS and landlocked developing countries, in accordance with their national plans and programmes
Goal 11 Make cities and human settlements inclusive, safe, resilient and sustainable	–
Goal 12 Ensure sustainable consumption and production patterns	–

SDG goals	SIDS-specific indicators
Goal 13 Take urgent action to combat climate change and its impacts	**13.b** Promote mechanisms for raising capacity for effective climate change-related planning and management in least developed countries and SIDS, including focusing on women, youth and local and marginalised communities
Goal 14 Conserve and sustainably use the oceans, seas and marine resources for sustainable development	**14.7** By 2030, increase the economic benefits to SIDS and least developed countries from the sustainable use of marine resources, including through sustainable management of fisheries, aquaculture and tourism **14.a** Increase scientific knowledge, develop research capacity and transfer marine technology, taking into account the Intergovernmental Oceanographic Commission Criteria and Guidelines on the Transfer of Marine Technology, in order to improve ocean health and to enhance the contribution of marine biodiversity to the development of developing countries, in particular SIDS and least developed countries
Goal 15 Protect, restore and promote sustainable use of terrestrial ecosystems, sustainably manage forests, combat desertification, and halt and reverse land degradation and halt biodiversity loss	—
Goal 16 Promote peaceful and inclusive societies for sustainable development, provide access to justice for all and build effective, accountable and inclusive institutions at all levels	—

(continued)

Table 1 (continued)

SDG goals	SIDS-specific indicators
Goal 17 Strengthen the means of implementation and revitalise the Global Partnership for Sustainable Development	17.18 By 2020, enhance capacity-building support to developing countries, including for least developed countries and SIDS, to increase significantly the availability of high-quality, timely and reliable data disaggregated by income, gender, age, race, ethnicity, migratory status, disability, geographic location and other characteristics relevant in national contexts

Source Author's own compilation based on UN (2015: 14–27)

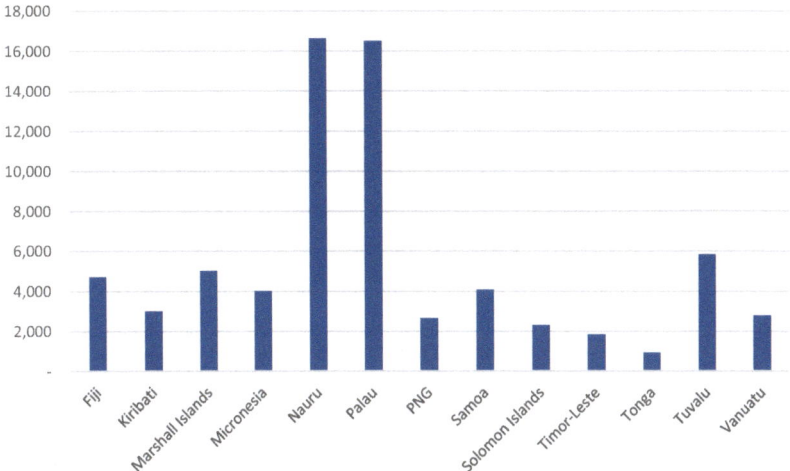

Fig. 1 Recent GNI per capita of Pacific SIDS (2019–2020, USD) (*Source* Author's own compilation based on World Bank data)

The remaining ten countries are MICs.[1] However, this does not mean that they are climate shock resilient, nor that they have achieved capacity to overcome sustainable development challenges.

As mentioned earlier, the SDGs, along with the UNFCCC, are a major global effort to achieve sustainable (and resilient) societies in countries. Accordingly, the UN High-Level Political Forum introduced the VNR process as part of the follow-up monitoring platform of the SDG implementation process. As all UN member countries are accountable for the SDGs, SIDS UN member countries are also submitting VNRs that present how SDGs are embedded in their national development plans. So far, 12 out of 13 Pacific SIDS UN members[2] have submitted VNRs, and only one country (Tuvalu) has not submitted its VNR, but is expected to

---

[1] According to World Bank (2021), LIC's GNI per capita is USD 1,045 or less, lower-middle income countries (LMIC)'s GNI per capita is between USD 1,046 and USD 4,095, upper-middle-income countries (UMIC)'s GNI per capita is USD 4,096 and USD 12,695 and HIC's GNI per capita is USD 12,696 or more.

[2] Fiji, Kiribati, Marshall Islands, Micronesia (Federated States of), Nauru, Palau, Papua New Guinea, Samoa, Solomon Islands, Timor-Leste, Tonga, Tuvalu and Vanuatu (UN SDGs Knowledge Platform 2021).

Table 2 SDGs VNR submission status of Pacific islands

Country	Submission year
Fiji	2019
Kiribati	2018
Marshall Islands	2021
Federated States of Micronesia	2020
Nauru	2019
Palau	2019
Papua New Guinea	2020
Samoa	2016, 2020
Solomon Islands	2020
Timor-Leste	2019
Tonga	2019
Tuvalu	To be submitted in 2022
Vanuatu	2019

Source Author's own compilation

submit it in 2022, as of August 2021. Table 2 shows when Pacific SIDS UN member countries submitted, or are expected to submit, their VNRs. As presented, Samoa has submitted its VNR twice, while the remaining Pacific UN member SIDS have done so once. This section intends to assess Pacific SIDS VNRs in order to examine how each government articulates their sustainable development plans, progress and challenges ahead, except Tuvalu whose VNR has not been submitted yet; and whether these Pacific SIDS government policies are commensurate with international concerns, as well as scholarship analysis, with regard to climate change in the context of the SDGs. Countries that submitted VNRs after 2019 managed to include the changing situation in light of the COVID-19 pandemic.

Fiji

Based on its mature democracy, Fiji has shown progress in areas such as low unemployment, universal access to primary education, decreased child and maternal mortalities, increased social welfare and economic services, and resilient and sustainable development against climate change and ocean degradation impacts. This progress was made based on Fiji's five-year and 20-year national development plans, along with the SDGs. The Fijian government prioritised development areas such as climate action, green growth, environmental protection, gender equality, disability assistance, good governance and economic and societal progress throughout the two plans. However, the government made it clear that Fiji requires more resources and capacity development in order to achieve the SDGs in their totality (Republic of Fiji 2019).

Kiribati

According to its VNR, Kiribati is one of the poorest countries and one of those most affected by climate change among Pacific SIDS. The country is experiencing all kinds of negative impacts of climate change. In terms of progress and challenges, it presents a somewhat mixed picture. For instance, while the government of Kiribati has achieved a remarkable increase in fishing revenues, its tuna fishery, which is the largest economic resource of the country, has been negatively affected by climate change. Education and employment have shown development, but not sufficient to address social and economic issues. The health sector remains vulnerable. With the development of the Kiribati Development Plan 2016–2019 and Kiribati Vision 20, the government has implemented the SDGs, but its institutional and financial capacities need to be improved (Government of Kiribati 2018).

Marshall Islands

The Republic of the Marshall Islands (RMI) has set out its National Strategic Plan in line with both the SDGs and the SAMOA Pathway. The government's plan presents five main priorities for development: social services and cultural identity; economic development; infrastructure; environmental awareness and climate change and good governance. With continuing disparities in the constituent islands' policies, the RMI government is pursuing a 'One Nation Concept' that can achieve a united and inclusive Marshallese identity, as well as social inclusion. In doing so, the government faces four main challenges: the negative impact of the COVID-19 pandemic; the country's small and isolated geography, and climate change impact; its high reliance on external resources and imports and underdeveloped human capital. In accordance with the SDGs, the Marshall Islands prioritises health, economic growth, inequality and climate change (Republic of the Marshall Islands 2021).

Federated States of Micronesia

The FSM prioritised eight sectors in relation to sustainable development: health; education; agriculture; fisheries; private sector development; transportation; communication; and, as a cross-cutting sector, energy. These priorities are in line with the country's national development strategy as contained in the FSM Strategic Development Plan 2004–2023, which emphasises sustainable economic growth and self-reliance. Even though the Federation has set out its priority areas to focus on development, it was inevitable that the government had to redirect resources to measures related to COVID-19. Also, the FSM's analysis shows that it lacks human resource capacity, and climate change and natural disasters have threatened its people's livelihood. Data analysis capacity in accordance with SDGs remains challenged too (Federated States of Micronesia 2020).

Nauru

Nauru is the smallest SIDS in the world, and thus, basically suffers from its small and remote location with limited production capacity. Nevertheless, the government of Nauru has achieved progress in the energy and economic sectors thanks to ODA. The government established the Nauru Sustainable Development Strategy in line with the SDGs and SAMOA Pathway; however, it turns out that massive external financial support is required, along with efforts to address lack of staff capacity and land capacity, as well as high NCD prevalence. On top of it, global economic shocks and climate change have made the country more vulnerable to resilience. In order to increase its level of adaptation, the most significant issue for the country is the relocation of communities and key infrastructure to higher ground, which has been a priority for the government of Nauru. However, relocation requires a restoration process, which, in turn, requires new and additional sources of funding (Government of the Republic of Nauru 2019).

Palau

Palau is one of the rare SIDS which has achieved HIC status. It was a LIC in 1994, but became a HIC in 2017. Accordingly, it achieved most of the MDGs, and now the government is striving for higher quality in most sectors. As such, the government of Palau has included all the SDGs in its development plan, under the four pillars of people, prosperity, planet and partnerships. However, as is the case with other SIDS, Palau is not protected from climate change, and thus, it is aiming to build a more resilient society against natural shocks (Republic of Palau 2019).

Papua New Guinea

Papua New Guinea (PNG) has delivered quite remarkable achievements in relation to the SDGs. For example, it has demonstrated progress towards achieving the poverty, hunger, health, education, energy, economic growth, climate change, land and partnership-related goals. While the country has received a high level of external financial support for climate resilient projects, both internal and external challenges have hindered progress towards meeting the goals of economic growth, consumption and production, peace, justice and institution and life below water, which have become the PNG government's development priorities for the next ten years (Papua New Guinea Government 2020).

Samoa

Samoa was one of the best economic performers in the region by 2008; however, as a SIDS, it could not overcome its weak resilience against natural disasters. Also, the global economic situation directly affected Samoa. The biggest manufacturing company in the country had to close, in 2017, due to the changing global market economy. Nevertheless, Samoa has become one of the few countries in the world to have successfully submitted its VNR more than once. Based on its experience of the previous VNR, the government of Samoa now focuses on the 'people goals' of the SDGs, with an emphasis on human capital and education. Samoa has been no exception to the COVID-19 pandemic and its impact. However, the country managed to obtain relatively high levels of ODA and foreign direct investment (FDI), which have sustained its efforts in building resilience and inclusive infrastructure, compared to other SIDS, amidst the pandemic. Land loss has affected biodiversity, and the country's remaining forest cover is non-native. It has been obvious that Samoa, though one of the strongest economies among Pacific SIDS, is still vulnerable to environmental shocks. In terms of the SDGs, one of the main challenges has been data analysis and management capacity (Government of Samoa 2020).

Solomon Islands

Solomon Islands has achieved remarkable development progress, espe-cially in the health and education sectors, which has led it to graduate from the list of least developed countries (LDCs) in 2024. However, the country still lags behind in resiliency against climate change in social and economic development. For instance, it has already lost five of its uninhabited islands due to sea-level rise, and six inhabited islands have been flooded. As the country is heavily dependent on forestry, the export of timber has resulted in a higher risk of flooding in local communi-ties exposed to stronger winds. Not only the rising sea level, but also the increased temperature has impacted on agriculture and fishing, which has led to insecurity in the country's food systems. The government of Solomon Islands also needs to empower civil servants and build stronger partnerships with civil society and the private sector in order to achieve sustainable development by 2030 (Solomon Islands Government 2020).

Timor-Leste

While achieving progress in health and education, the government of Timor-Leste has set out a Strategic Development Plan that focuses on three main pillars to achieve sustainable development on the island, with priority placed on human capital: state-building; social inclusion and economic growth. Timor-Leste has also managed to improve its national statistical capacity; however, it struggles to produce and analyse timely and reliable high-quality data. At the same time, the country lacks international financial support, including private sector involvement, and is in significant need of decent jobs for its young men and women (Government of Timor-Leste 2019).

Tonga

Tonga aims to achieve sustainable and inclusive growth in line with the SDGs, SAMOA Pathway, Addis Ababa Agreement, and Sendai Frame-work for Disaster Risk Reduction, built upon the Tonga Strategic Development Framework 2015–2025. Accordingly, it prioritises social protection and human rights, focusing on vulnerable groups. It also emphasises its continuing commitment to Universal Health Coverage, by considering domestic health-related concerns such as its serious NCD burden. The government of Tonga has also addressed the importance of education, energy, economic growth, equality, climate change, justice and institutions and partnership in the context of the SDGs. The government has recognised the need for enhanced public sector management, including national development planning systems (Kingdom of Tonga 2019).

Vanuatu

The government of Vanuatu clearly defines itself as a SIDS whose small land-size and limited resources make it more vulnerable to environmental challenges and global economic shocks. Nevertheless, Vanuatu has been the first Pacific country to develop a National Implementation Plan for the Universal Periodic Review recommendations. Also, it has established Vanuatu 2030—a national sustainable development plan in line with the SDGs. According to the Vanuatu government, its development priorities are inclusivity and equality, with a focus on education, economic growth, inequality, climate change, peace and global partnership for achieving the SDGs. With the aim of becoming a 'global leader in disaster recovery', the government has developed the National Disaster Recovery Framework as well as the National Policy on Climate Change and Disaster Risk Reduction 2016–2030 (Republic of Vanuatu 2019).

As shown, climate change has been included in all the SIDS narratives; however, it is not at the core of each country's concerns in terms of sustainable development. While all the countries tend to accept that climate change and extreme weather events are challenges and obstacles, they also tend to emphasise economic growth and social development, including human resource development and education. Some countries consider private sector involvement to be important for development, and other countries identify data management capacity as one of the main difficulties. Yet others emphasise their continuing need for external financial support in order to become resilient societies, regardless of their economic status in terms of GNI.

Donor Support of Pacific Sustainable Development: The Case of Australia

Among SIDS, the Pacific islands in particular are highly dependent on ODA for their national budgets (Dornan and Pryke 2017). According to

Wood (2020: 115), Pacific island countries comprise nine of the world's 15 most aid-dependent recipients, with development aid amounting to about 50% of their economies. Among them, as seen in Fig. 2, Papua New Guinea is by far the largest recipient in the region, with most of its aid received from Australia, followed at a huge distance by French Polynesia and New Caledonia. Interestingly, neither French Polynesia nor New Caledonia has received any ODA since 2000, while the Marshall Islands and Micronesia began to receive ODA from 1990 onwards.

Among donors, Australia is the leading OECD DAC donor in the region, while China is the leading donor among non-DAC countries. However, in terms of amounts, Australia provides far more aid to the Pacific islands than China (Wood 2020). Figure 3 depicts trends in aid from the main donors to the Pacific region over the most recent five-year period. The OECD ODA dataset includes only DAC member countries, with donors like China not included in the OECD ODA statistics. Thus, Fig. 3 has been compiled based on the Lowy Institute Pacific Aid Map, which includes all donor data for the Pacific region, regardless of OECD

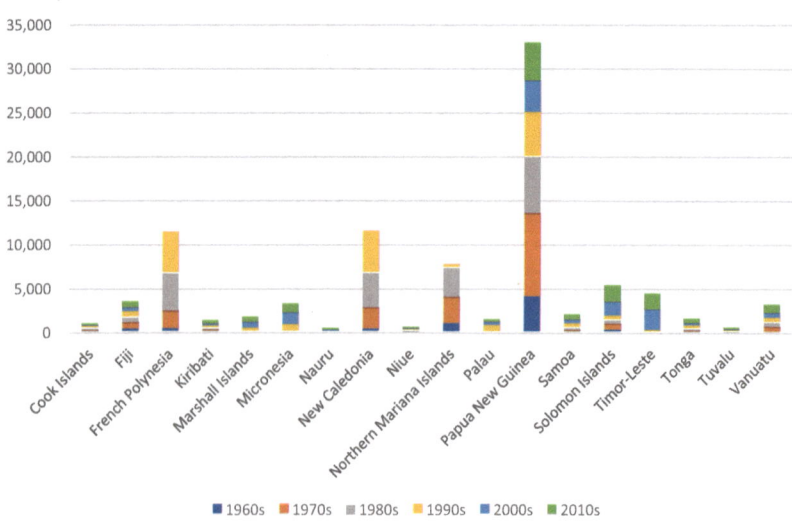

Fig. 2 ODA trends of the Pacific SIDS (Disbursement, USD million) (*Source* Author's own compilation based on OECD Statistics. https://stats.oecd.org/ Index.aspx?ThemeTreeID=3&lang=en#)

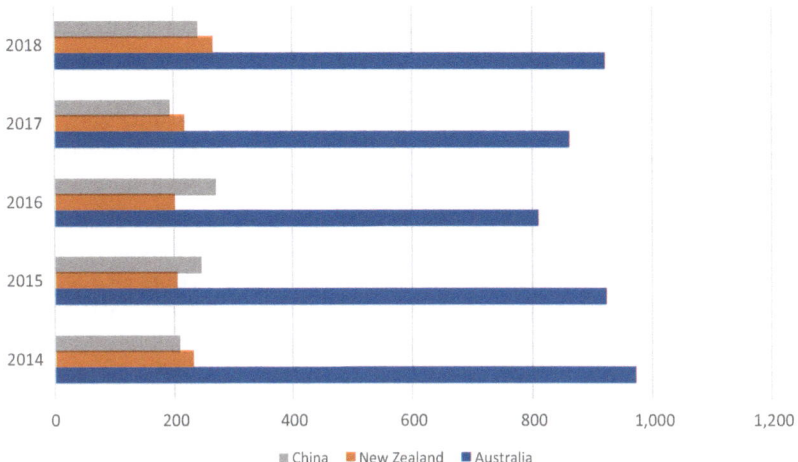

Fig. 3 Main Donor Aid trends in the Pacific region (Disbursement, USD million) (*Source* Author's own compilation based on Lowy Institute Pacific Aid Map. https://pacificaidmap.lowyinstitute.org/)

DAC membership. The Lowy Institute data for OECD DAC members differs slightly from that of the OECD in absolute terms; however, the trends they show are identical in that Australia is the largest donor, followed at some distance by New Zealand and China.

The government of Australia has provided development aid to 12 Pacific states, 11 of which have signed Aid Partnership Arrangements with Australia (DFAT 2021). While the UN has categorised Timor-Leste as one of 13 Pacific SIDS, Australia does not include Timor-Leste in the Pacific islands category. Accordingly, the aforementioned 12 Pacific states refer to the SIDS analysed in the previous section, except Timor-Leste. As the Australian government pursues an integrated approach to its foreign, trade, security and development policies, it merged the country's aid agency, AusAID, into the Department of Foreign Affairs and Trade (DFAT) in 2013 (DFAT 2015e; Grattan 2013) and established its development aid policy of 'Australian Aid: Promoting Prosperity, Reducing Poverty and Enhancing Stability'. This policy serves as the overarching policy for all its ODA to Pacific recipients.

Based on this Australian Aid policy, the government provides detailed country-specific policies in the form of DFAT Aid Investment Plans

(AIPs), which were published in 2015. The AIPs were written based on discussions between Australia and each recipient government (for example, see DFAT 2015b). While DFAT delivers its aid based on the AIPs, these AIPs do not extensively reflect the SDGs, because they were developed in 2015—the SDGs were agreed at the UN General Assembly one month after the AIPs were published. Nevertheless, it would be logical to reason that the Australian government has reflected the SDGs into its current workstream, as the AIPs considered progress towards the MDGs, which are predecessors of the SDGs. As a matter of fact, the VNR that the Australian government submitted in 2018 clearly shows that its aid programmes are in line with the sustainable development agenda in the Pacific region (Australian Government 2018). In terms of Australia's aid to Timor-Leste, the government aid policy can be found in the East Timor Strategic Planning Agreement for Development. In order to understand how Australia's aid policy reflects the needs of each aid recipient government in the Pacific region, each country AIP has been examined below. In the case of Timor-Leste, the agreement document has been analysed.

Fiji

While Fiji has been an important partner for Australia, Australia has been the largest bilateral donor to Fiji. With a relatively high development level, Fiji tends to focus on private sector development, and this has been well addressed in Australian development aid policy. In other words, the government of Australia recognises the Fijian government's development priorities of private sector growth and an economy with higher employment, based on Fiji's 2015 National Development Plan. Accordingly, Australia has jointly identified development priorities in Fiji and set two strategic priorities in its development aid: increased private sector development; and improved human development. Not only the country's vulnerability to climate change, but also good governance and gender equality have been the other main areas of focus (DFAT 2015b).

Kiribati

According to the government of Australia, Kiribati is one of the poorest states in the Pacific, and Australia has become a leading donor in this country. The 2012–2015 Kiribati Development Plan focused on economic development and labour mobility, which Australia has reflected into its aid policy towards Kiribati. Accordingly, Australian aid policy towards Kiribati serves two priority areas: economic reform; and a better educated and healthier population. In doing so, there has been a strong emphasis on fishery as Kiribati heavily relies on fisheries revenue. Also, gender equality, disability inclusiveness and environment resilience have been the main areas for improvement in terms of capacity (DFAT 2015c).

Marshall Islands

The Australian government locates the RMI as a SIDS in the Pacific region and has addressed the island country's high level of vulnerability to climate change and natural disasters, along with its fragile economy. The main source of aid for the Marshall Islands has been the US Compact of Free Association grant scheme, which constitutes the RMI Compact Trust Fund; however, the scheme ends in 2023. Australia has reflected the Marshallese government's economic and social reform process, which was embedded in RMI National Strategic Development Framework Vision 2018 and National Strategic Plan 2015–2017, in its aid policy. Australia has set out two main objectives in its ODA programme for RMI: clean water and sanitation in Ebeye island; and gender equality and women's empowerment. DFAT especially works with the Asian Development Bank (ADB) in providing support to the Marshall Islands. Also, Australia supports RMI's Paris Agreement implementation process as part of its climate support (DFAT 2015h).

Federated States of Micronesia

The FSM mainly receive aid from the US Compact grant, the FSM Compact Trust Fund, which terminates in 2023. Australia identifies FSM as SIDS and works with the ADB to support the island country's education sector. Even though FSM's main revenue comes from fishery, DFAT tends to focus more on other development areas by aligning with priorities in the FSM Strategic Development Plan 2023: quality and inclusive primary education; and gender equality and women's empowerment. It also contributes to climate support by providing assistance for FSM's Paris Agreement implementation efforts (DFAT 2015a).

Nauru

DFAT reflects Nauru's National Sustainable Development Strategy into its development aid policy towards Nauru. Nauru requires long-term development assistance, as it experiences not only capacity deficiency in human development and economic performance, but also challenges such as geographic isolation and heavy dependency on fisheries. Based on this, the Australian government provides for three main strategic priorities: more effective public sector management; nation-building infrastructure and human development (DFAT 2015d).

Palau

Australia has recognised Palau as a SIDS; however, its population's standard of living has been assessed as one of the highest in the Pacific region due to tourism income, private sector engagement and US assistance. Palau has benefitted from the Palau Compact Trust Fund, which remains in place until 2024, provided by the US Compact grant. While addressing the 2020 Palau National Master Development Plan, the government of Australia has set out two main strategic focus areas in its development aid to Palau: economic growth through upgraded telecommunications and internet coverage; and gender equality and women's empowerment. Accordingly, DFAT gives special attention to reform of the information and communications technology (ICT) sector in Palau. The government of Australia also supports Palau's efforts to implement the Paris Agreement in order to achieve climate resilience (DFAT 2015f).

Papua New Guinea

Australia has been the largest donor to PNG, with which it has a long partnership. PNG has achieved strong economic growth and aims to achieve upper-middle-income country (UMIC) status by 2050. However, natural disasters have been a serious challenge to this ambition, and thus DFAT focuses its aid assistance on this issue. The three main pillars of Australia's aid programme for PNG are aligned with the PNG government's priorities in, for example, the 2016–2017 Medium Term Development Plan 2: promoting effective governance; enabling economic growth and enhancing human development. These are integrated with gender equality as a cross-cutting issue (DFAT 2015g).

Samoa

Samoa has achieved sound institutional capacity, including stable parliamentary democracy; however, it still needs ODA for its economic growth as it suffers from climate change impacts, such as natural disasters and NCDs. Accordingly, DFAT has set out three main areas to address in its aid policy: enabling economic growth; progressing health and education outcomes and strengthening governance. In Samoa, the aforementioned three priorities reflect Samoa's Strategy for the Development of Samoa and Development Cooperation Policy (DFAT 2015i).

Solomon Islands

Australia has been the largest donor to the Solomon Islands. DFAT's bilateral aid to the Solomon Islands is designed by addressing the priorities provided in the Solomon Islands government's National Development Strategy and Medium Term Development Plan. In spite of its achievements in economic development, the Solomon Islands has experienced economic reversals due to conflict and natural disasters, and become fragile. With this in mind, DFAT has set out three main aid objectives for the Solomon Islands: supporting stability; enabling economic growth and enhancing human development. While Australian bilateral aid to the Solomon Islands is channelled through the Regional Assistance Mission to Solomon Islands (RAMSI), its multilateral aid is provided through the World Bank, ADB and other UN bodies in the Solomon Islands. DFAT also works actively with non-governmental organisations (NGOs) (DFAT 2015j).

Timor-Leste

The Australian government considers the Dili Development Pact, which was built upon Timor-Leste's Strategic Development Plan 2011–2030. Based on this, Australia pursues four priority areas in its aid support to Timor-Leste: sustainable economic growth; increasing access to quality education; increasing access to quality health services and to safe water and sanitation and effective governance (DFAT 2011). As the Strategic Plan is written in detail, the agreement document, which was mentioned above, itself does not provide much information (see Government of Timor-Leste 2011).

Tonga

Even though Tonga has maintained a relatively high level of political and social stability, including its maturing parliamentary democracy, it is at risk of natural disasters and lacks resilience capacity. For example, it has been assessed as being the 'third highest risk globally (after Vanuatu and Philippines) of disaster' according to the AIP (DFAT 2015k: 2). Also, the country still requires ODA for its economic development, along with private sector engagement. Even though Australia's support has tackled NCD prevalence in the country, it is still one of the main challenges to Tonga's development. In light of this, DFAT has established three strategic priorities in line with the Tongan government's development strategy: governance, economic and private sector development reforms; a more effective, efficient and equitable health system and skills development in support of economic opportunities for Tongan workers (DFAT 2015k).

Tuvalu

While Tuvalu has not published its VNR at the time of writing, the government has provided its own development plan, Te Kakeega Ill: National Strategy for Sustainable Development 2016–2020, based on the SDGs and SAMOA Pathway. Accordingly, the government of Australia reflects this into its aid programme while focusing on the capacity development of Tuvalu, so that it can use its revenue and human resources for development. Also, NCDs and climate change are emphasised in Australis's aid programme for Tuvalu. Overall, DFAT focuses on three main areas of aid support in Tuvalu: economic and financial management systems; basic service delivery in the education and health sectors and climate change adaptation and disaster preparedness and response (DFAT 2015l).

Vanuatu

The Australian government supports Vanuatu based on shared priority areas by reflecting the priorities in Vanuatu's National Sustainable Development Plan and the National Recovery and Economic Strengthening Programme Plan—Strengthening ni-Vanuatu Resilience. In doing so, DFAT's aid support prioritises four main development objectives: resilient

infrastructure and environment for economic opportunity; early education and essential health services; community safety and resilience and
cyclone recovery and reconstruction. Although Vanuatu has achieved relatively strong economic growth in the region, with financial support from
the Australian government, it is still highly vulnerable to climate change
impact. For example, its city of Port Vila is recorded as the 'world's most
exposed city' to natural disasters, while the country itself has been noted
as the 'world's most vulnerable country' to extreme natural events (DAFT
2015m: 3). Thus, the government of Australia provides aid to areas such
as climate change and disaster recovery, along with gender equality and
disability inclusion (DFAT 2015m).

Overall, the government of Australia has addressed what each recipient
government in the Pacific region faces in terms of development challenges. Apart from bilateral assistance, Australia focuses on four main areas
in its Pacific regional programme: economic growth; effective regional
institutions; healthy and resilient communities and empowering women
and girls. For instance, DFAT contributes 20% of its Pacific aid budget
to regional institutions, such as the Green Climate Fund (GCF) and
the University of the South Pacific (DFAT 2015e: 2 and 12). The
government of Australia recognises the difficulties associated with climate
change, including natural disasters, in the Pacific islands' development
pathway, while its aid programme is designed based on its development
policy of 'Australian Aid: Promoting Prosperity, Reducing Poverty and
Enhancing Stability' (DFAT 2015e: 3).

Conclusion

This chapter has attempted to examine linkages and disparities among
SIDS climate change theories, global sustainable development narratives,
the Pacific islands' climate adaptation and mitigation efforts in the context
of their SDG implementation policies and Australian aid policies. It has
explored the climate change and SIDS discussion, and how this has been
presented in the contemporary global norm, UN Sustainable Development Agenda, including SDGs. And it has analysed the case of Pacific
SIDS, focusing on UN SDGs VNRs, along with the case of Australia
as donor. In theory, SIDS are major victims of climate change, and
their resilience capacity is vulnerable. One of the most concerning consequences of climate change is sea-level rise, as small islands are becoming
even smaller, which makes their land capacity for habitat and economic

activities fragile. Academics specifically discuss secondary problems from climate change effects. However, not all cases have led to distressing situations; climate adaptation efforts over the last 30 years have shown promising successful cases too. When the term SIDS was created, the focus was on smallness; however, thanks to encouraging adaptation cases, smallness no longer means core vulnerability. Yet, it is also plain that repeating extreme weather shocks are clear obstacles for sustainability in these countries, and thus donor support for customised adaptation and resilience capacity development is crucial. For example, SIDS were left behind, as mentioned earlier, in MDG implementation, and thus, they have been categorised as the most vulnerable county-group.

Given this, the SDGs have specifically provided indicators for SIDS, including human resource development and resilient infrastructure development, along with more ODA flows. However, the international community did not articulate SIDS-specific targets for goals like food security and nutrition (SDG 2) and human settlement and resilient and sustainable cities (SDG 11). It is surprising that these goals do not include SIDS-specific indicators, given the academic discussion on how food insecurity and diseases like NCDs have been the main issues facing SIDS affected by climate change. Also, due to their coastal and small land characteristics, resilient cities and human settlement issues cannot be emphasised enough for SIDS. Article 2.4 (indicator of SDG 2) of the Sustainable Development Agenda document states: 'By 2030, ensure sustainable food production systems and implement resilient agricultural practices that increase productivity and production, that help maintain ecosystems, that strengthen capacity for adaptation to climate change, extreme weather, drought, flooding and other disasters and that progressively improve land and soil quality' (UN 2015: 15). Also, Article 11.b (indicator of SDG 11) states: 'By 2020, substantially increase the number of cities and human settlements adopting and implementing integrated policies and plans towards inclusion, resource efficiency, mitigation and adaptation to climate change, resilience to disasters, and develop and implement, in line with the Sendai Framework for Disaster Risk Reduction 2015–2030, holistic disaster risk management at all levels' (UN 2015: 22). Both address climate change adaptation issues in food security and habitat-related areas. However, as pointed out at the beginning of the document as well as by scholars, these kinds of indicators are more relevant and important to SIDS in order for them not to be left behind in obtaining the necessary external support.

Accordingly, development priorities under the SDG mechanism for Pacific SIDS include areas such as gender equality (SDG 5, which does not have SIDS-specific indicators). Also, food security, health issues such as NCDs, and land capacity for inhabited islands are among the most frequently repeated development issues in Pacific SIDS VNRs, along with their need for more financial support. While most countries want to achieve economic growth and resilience, the role of the private sector has also been frequently mentioned. This tells us that there is a gap between theories, global policies and national policies to some extent. At the same time, it is somewhat unclear as to whether creative and/or native local adaptation efforts have been recognised by each government in its development path. While existing research argues that local voices can increase the likelihood of adaptation and thus need to be included in national policies, it seems that such small details have not been recognised in the small islands yet, even though these can become an important game changer.

A major donor—Australia—also does not seem to be fully in line with theories, SDGs and each recipient's narratives. For example, while the Fijian government has set priorities including climate action and environmental protection, gender equality and economic and societal progress, Australian aid policy to Fiji addresses private sector development and human development. This does not mean that the two issues highlighted by the Australian government are not commensurate with Fiji's development plans; rather, it is that the Australian government has reconciled Fiji's priorities with its overarching aid policies. The case of Kiribati provides another example. Kiribati, in its VNR, pointed out the vulnerability of the health sector, and Australia includes building a healthier population as one of its two priorities. However, other areas of emphasis or weak areas of development have not been identical. Of course, we cannot ignore the possibility that these disparities have occurred due to the difference in timing between when the Australian aid policy document was written and when each recipient's VNR was written. Nevertheless, it seems that priorities raised in the same recipient's national development plan have been differently interpreted. When it comes to climate change issues, compliance with global norms on the environment has been clearly included in Australia's aid policy on Pacific SIDS. This is evident in that aid finance is channelled to climate mitigation efforts. However, it is somewhat unclear as to whether and how the donor government recognises locally developed adaptation efforts.

As seen, what we can observe from existing scholarly studies and aid recipient countries development plans is that the impact of climate change on SIDS has not been sufficiently linked to the UN Sustainable Development Agenda. Even though UN members have recognised the differentiated situations in SIDS and the issue of inequality in confronting climate change, the specific consequences that need to be addressed have not been considered. This then brings us to a somewhat puzzling picture. On one hand, while the SDGs provide certain indicators specifically aimed at SIDS, with both SIDS and donors expected to give more attention to the achievement of these indicators, the Pacific small islands have not given more weight to these indicators, but prioritised those issues that have appeared in scholarly discussions. On the other hand, however, the UN Sustainable Development Agenda has recognised the need to enhance the statistical capacities of SIDS (see UN 2015: 12 and 32; see also Table 1), by stating that international financial support and debt relief are important for SIDS capacity to achieve sustainable development (UN 2015: 11 and 29; see also Table 1). Accordingly, both recipients and donors tend to emphasise data management capacity.

It can be argued that economic and social development has deteriorated due to the consequences of climate disasters and thus climate change mitigation should be put ahead of adaptation. However, as it has been pointed out, islanders can develop better adaptation methods based on their differentiated situations and their own experiences. Climate change is not an issue which only SIDS can mitigate, but rather one which requires action at the global level. However, adaptation can be boosted better when it is associated with local experience and indigenous knowledge. Also, as emphasised, small islanders have been major victims of the effects of global warming. Most importantly, the rhetoric of SIDS being more seriously impacted by climate change is not new; it has been around for the last 30 years. But they are still vulnerable in terms of resilience, with differences in each country's need in the context of the SDGs. The solutions look small for the international community, but are not small for the small islanders. In order not to leave SIDS behind again under the SDGs mechanism, donor financial resources need to be targeted at locally tailored adaptation methods and improving their adaptation capacity. Economic and social development are equally important in SIDS, but the constant direct and indirect consequences of climate change and thus lack of resilience capacity overshadow the common developmental needs.

As the recent COVID-19 pandemic has hit SIDS, without exception, it may be reshuffling the landscape of what they have achieved so far. Some countries' progress could have been affected harder than that of others. However, this chapter does not intend to deal with COVID-19-related analysis yet; Australia's pre-emptive COVID-19 strategy in the region, which has been set out as a two-year plan (DFAT 2020), is not highly relevant to the SDG context. In light of this, COVID-19-related research and analysis could be carried out as a follow-up to this study. In addition, further research that includes donors such as China could be conducted for a comparative study. For example, follow-up research could compare Australian (OECD DAC) and Chinese (non-DAC) aid policies towards Pacific SIDS in terms of theory, the global norm and recipient policies in the context of the SDGs and climate change. This would contribute to further discussion on linkages and disparities among SIDS climate change theories, global norms, recipients' individual challenges and donor aid policy in the context of the SDGs, as a way of bridging the gap between policy and reality.

References

Australian Government (2018). Report on the Implementation of the Sustainable Development Goals. Department of Foreign Affairs and Trade.

DFAT (2011). Strategic Planning Agreement for Development between the Government of Timor-Leste and the Government of Australia. https://www.dfat.gov.au/about-us/publications/Pages/east-timor-strategic-planning-agreement-for-development-english. Accessed: 28 July 2021.

DFAT (2015a). Aid Investment Plan: Federated States of Micronesia 2016–17 to 2018–19. https://www.dfat.gov.au/about-us/publications/Pages/aid-investment-plan-aip-fsm-2016-17-to-2018-19. Accessed: 28 July 2021.

DFAT (2015b). Aid Investment Plan: Fiji 2015–16 to 2018–19. https://www.dfat.gov.au/about-us/publications/Pages/aid-investment-plan-aip-fiji-2015-16-to-2018-19. Accessed: 28 July 2021.

DFAT (2015c). Aid Investment Plan: Kiribati 2015–16 to 2018–19. https://www.dfat.gov.au/about-us/publications/Pages/aid-investment-plan-aip-kiribati-2015-16-to-2018-19. Accessed: 28 July 2021.

DFAT (2015d). Aid Investment Plan: Nauru 2015–16 to 2018–19. https://www.dfat.gov.au/about-us/publications/Pages/aid-investment-plan-nauru-2015-16-to-2018-19. Accessed: 28 July 2021.

DFAT (2015e). Aid Investment Plan: Pacific Regional 2015–16 to 2018–19. https://www.dfat.gov.au/about-us/publications/Pages/aid-investment-plan-aip-pacific-regional-2015-16-to-2018-19. Accessed: 28 July 2021.

DFAT (2015f). Aid Investment Plan: Palau 2016–17 to 2018–19. https://www.dfat.gov.au/about-us/publications/Pages/aid-investment-plan-aip-palau-2016-17-to-2018-19. Accessed: 28 July 2021.

DFAT (2015g). Aid Investment Plan: Papua New Guinea 2015–16 to 2017–18 (extended to 2018–2019). https://www.dfat.gov.au/about-us/publications/Pages/aid-investment-plan-aip-papua-new-guinea-2015-16-to-2017-18. Accessed: 28 July 2021.

DFAT (2015h). Aid Investment Plan: Republic of the Marshall Islands 2016–17 to 2018–19. https://www.dfat.gov.au/about-us/publications/Pages/aid-investment-plan-aip-marshall-islands-2016-17-to-2018-19. Accessed: 28 July 2021.

DFAT (2015i). Aid Investment Plan: Samoa 2015–16 to 2018–19. https://www.dfat.gov.au/about-us/publications/Pages/aid-investment-plan-aip-samoa-2015-16-to-2018-19. Accessed: 28 July 2021.

DFAT (2015j). Aid Investment Plan: Solomon Islands 2015–16 to 2018–19. https://www.dfat.gov.au/about-us/publications/Pages/aid-investment-plan-aip-solomon-islands-2015-16-to-2018-19. Accessed: 28 July 2021.

DFAT (2015k). Aid Investment Plan: Tonga 2015–16 to 2018–19. https://www.dfat.gov.au/about-us/publications/Pages/aid-investment-plan-aip-tonga-2015-16-to-2018-19. Accessed: 28 July 2021.

DFAT (2015l). Aid Investment Plan: Tuvalu 2016–17 to 2019–20. https://www.dfat.gov.au/about-us/publications/Pages/aid-investment-plan-aip-tuvalu-2016-17-to-2019-20. Accessed: 28 July 2021.

DFAT (2015m). Aid Investment Plan: Vanuatu 2015–16 to 2018–19. https://www.dfat.gov.au/about-us/publications/Pages/aid-investment-plan-aip-vanuatu-2015-16-to-2018-19. Accessed: 28 July 2021.

DFAT (2020). Partnership for Recovery: Australia's COVID-19 Development Response.

DFAT (2021). Overview of Australia's Aid Programme to the Pacific: Enduring Partners. https://www.dfat.gov.au/geo/pacific/development-assistance/Pages/overview-of-australias-aid-program-to-the-pacific. Accessed: 28 July 2021.

Donner, Simon D. (2015). The Legacy of Migration in Response to Climate Stress: Learning from the Gilbertese Resettlement in the Solomon Islands. *Natural Resources Forum*, 39: 191–201.

Dornan, Matthew and Jonathan Pryke (2017). Foreign Aid to the Pacific: Trends and Developments in the Twenty-First Century. *Asia & the Pacific policy Studies*, 4(3): 386–404.

Federated States of Micronesia (2020). Our Actions Today Are Our Prosperity Tomorrow: First Voluntary national Review on the 2030 Agenda for Sustainable Development. Pohnpei, Federated States of Micronesia.

Gibson, K. E., J. Barnett, N. Haslam, and I. Kaplan (2020). The Mental Health Impacts of Climate Change: Findings from a Pacific Island Atoll Nation. *Journal of Anxiety Disorders*, 73: 1–8.

Gohar, Abdelaziz A., Adrian Cashman, and Frank A. Ward (2019). Managing Food and Water Security in Small Island States: New Evidence from Economic Modelling of Climate Stressed Groundwater Resources. *Journal of Hydrology*, 569: 239–251.

Government of Kiribati (2018). Voluntary National Review and Development Plan Mid-Term Review.

Government of Samoa (2020). Second Voluntary National Review on the Implementation of the Sustainable Development Goals to Ensure Improved Quality of Life for All. Apia, Ministry of Foreign Affairs and Trade.

Government of the Republic of Nauru (2019). Voluntary National Review on the Implementation of the 2030 Agenda.

Government of Timor-Leste (2011). Timore-Leste Strategic Development Plan 2011–2030: Version Submitted to the National Parliament.

Government of Timor-Leste (2019). Report on the Implementation of the sustainable Development Goals: From Ashes to Reconciliation, Reconstruction and Sustainable Development. Voluntary National Review of Timor-Leste 2019. Dili, Timor-Leste.

Grattan, Michelle (2013). DFAT Secretary's Tough Message about AusAID Integration. Conversation, 3 November 2013. https://theconversation.com/dfat-secretarys-tough-message-about-ausaid-integration-19799. Accessed: 28 July 2021.

Halsted, Erin (2016). Citizens of Sinking Islands: Early Victims of Climate Change. *Indiana Journal of Global Legal Studies*, 23(2): 819–837.

Kelman, Ilan (2010). Hearing Local Voices from Small Island Developing States for Climate Change. *Local Environment*, 15(7): 605–619.

Kelman, Ilan (2018). Islandness Within Climate Change Narratives of Small Island Development States (SIDS). *Island Studies Journal*, 13(1): 149–166.

Kelman, Ilan, Sonja Ayeb-Karlsson, Kelly Rose-Clarke, Audrey Prost, Espen Ronneberg, Nicola Wheeler, and Nicholas Watts (2021). A Review of Mental Health and Wellbeing under Climate Change in Small Island Developing States (ISDS). *Environmental Research Letters*, 16: 1–13.

Kingdom of Tonga (2019). Voluntary National Review. Nuku'alofa, Prime Minister's Office.

Lim, Sojin (2016). Recent Trends and Issues of the International Development Cooperation. In: KOICA ODA Education Centre (ed). International Development Cooperation. Seongnam, Sigong Media. [임소진 (2016). 국제개발협

력 최근 동향과 이슈. KOICA ODA 교육원 (편집). 국제개발협력 입문편 중. 성남, 시공미디어].

Lowitt, Kristen, Arlette Saint Ville, Patsy Lewis, and Gordon M. Hickey (2015). Environmental Change and Food Security: The Special Case of Small Island Developing States. Reg Environ Change, 15: 1293–1298.

Mclver, Lachian, Rokho Kim, Alistair Woodward, Simon Hales, Jeffery Spickett, Dianne Katscherian, Masahiro Hashizume, Yasushi Honda, Ho Kim, Steven Iddings, Jyotishma Naicker, Hilary Bambrick, Anthony J. McMichael, and Kristie L. Ebi (2016). Health Impacts of Climate Change in Pacific Island Countries: A Regional Assessment of Vulnerabilities and Adaptation Priorities. *Environmental Health Perspectives*, 124(11): 1707–1714.

OECD (2021). Making Development Cooperation Work for Small Island Developing States. Paris, OECD.

Papua New Guinea Government (2020). Progress of Implementing the Sustainable Development Goals: Papua New Guinea's Voluntary National Review. Port Moresby, Department of National Planning and Monitoring.

Perumal, Nikita (2018). "The Place Where I Live is Where I Belong": Community Perspectives on Climate Change and Climate-Related Migration in the Pacific Island Nation of Vanuatu. *Island Studies Journal*, 13(1): 45–64.

Republic of Fiji (2019). Voluntary National Review: Fiji's Progress in the Implementation of the Sustainable Development Goals. Suva, Ministry of Economy.

Republic of Palau (2019). Pathway to 2030: Progressing with our Past Toward a Resilient, Sustainable and Equitable Future. 1st Voluntary National Review on the SDGs.

Republic of the Marshall Islands (2021). Voluntary National Review.

Republic of Vanuatu (2019). Voluntary National Review on the Implementation of the 2030 Agenda for Sustainable Development.

Rudiak-Gould, Peter (2014). Climate Change and Accusation: Global Warming and Local Blame in a Small Island State. *Current Anthropology*, 55(4): 365–386.

Savage, Amy, Lachlan Mclver, and Lisa Schubert (2020). Review: The Nexus of Climate Change, Food and Nutrition Security and Diet-Related Non-Communicable Diseases in Pacific Island Countries and Territories. *Climate and Development*, 12(2): 120–133.

Scandurra, G., A. A. Romano, M. Ronghi, and A. Carfora (2018). On the Vulnerability of Small Island Developing States: A Dynamic Analysis. *Ecological Indicators*, 83: 382–392.

Solomon Islands Government (2020). Solomon Islands Voluntary National Review.

Thomas, Adelle, April Baptiste, Rosanne Martyr-Koller, Patrick Pringle, and Kevon Rhiney (2020). Climate Change and Small Island Development States. *Annual Review of Environment and Resources*, 45: 6.1–6.27.

UN (2015). Transforming Out World: the 2030 Agenda for Sustainable Development. Resolution Adopted by the General Assembly on 25 September 2015. A/RES/70/1. New York, United Nations.

UNDP (2017). Financing the SDGs in the Pacific islands: Opportunities, Challenges and Ways Forward. New York, United Nations Development Programme.

UN SDGs Knowledge Platform (2021). Small Island Developing States. https://sustainabledevelopment.un.org/topics/sids/list. Accessed: 28 July 2021.

Walshe, Rory A., and Charlotte Eloise Stancioff (2018). Small Island Perspectives on Climate Change. *Island Studies Journal*, 13(1): 13–24.

Wood, Terence (2020). COVID-19 and Australian Aid in the Pacific. East Asia Forum, 19 August 2020. https://www.eastasiaforum.org/2020/08/19/covid-19-and-australian-aid-in-the-pacific/. Accessed: 28 July 2021.

World Bank (2021). World Bank Country and Lending Groups. https://datahelpdesk.worldbank.org/knowledgebase/articles/906519-world-bank-country-and-lending-groups. Accessed: 29 July 2021.

New Zealand's Political Responses to Climate Change and Migration in the Pacific: A Perspective from the South

Ti-han Chang⦿ *and Lyn Collie*⦿

Today, climate displacement is considered a pressing political issue, worthy of urgent address. In the South Pacific region particularly, there has been a consistent shaping of climate migration narratives since 2000. In this regional context, climate migration has so far mainly been framed and addressed in the theoretical or political paradigm of climate justice. However, climate justice aligns itself conceptually with a Euro-/American-centric perspective (or, speaking more reductively, the so-called "Western" position).[1] Our chapter attempts to push past this

[1] Looking at how the concept of climate justice first developed, it is evident that it is an extension of the concept of environmental justice. In the 1970s and 1980s, when the concept of environmental justice first emerged in the United States, it focussed on the NIMBY-ism (Not In My Back Yard), which associated racial/class discrimination and local environmental pollution. It is a concept that originated in the U.S. and thus its political development follows a logic inherent to Euro-/American-centric values.

T. Chang (✉)
University of Central Lancashire, Preston, UK
e-mail: TChang2@uclan.ac.uk

L. Collie
Auckland, New Zealand

framework, examining the impact of New Zealand and Kiribati's existing climate migration policies and positions, to argue that solutions to climate displacement should be reconsidered in their regional context. In Kiribati, there has previously been a focus on how to mitigate the impact of climate change by encouraging intergovernmental policies which channel resources into labour migration, mostly relocating I-Kiribati people to their developed neighbours (particularly New Zealand and Australia). However, we recognise that this approach assumes a Eurocentric set of values based on the principles of human rights and free labour mobility and produces a number of negative social outcomes for migrants. In this chapter, we examine some of the social and environmental outcomes of migration and adaptation policy in New Zealand and Kiribati, concluding that the narrative of climate justice becomes problematic when used in the South Pacific to develop policy responses to climate migration.

To address the issue in more depth, this chapter is divided into four sections. In the first section, we examine New Zealand's current struggles with racial equality, including for Pacific peoples, in the context of its immigration history. We suggest that, while New Zealand can be considered socially progressive in many respects, the purposeful creation of a British hegemony in colonial New Zealand has formed the basis of enduring institutional and structural racism,[2] including against Pacific people. We explore how contemporary Pacific migrants, coming into the social and political context of New Zealand, become part of an existing under-privileged community when they relocate. Second, our analysis draws attention to census data and studies which reveal more about recent Pacific migrants' social outcomes and lived experiences in New Zealand, alongside the specific experiences of I-Kiribati migrants.[3] Our analysis demonstrates that social, cultural, and economic discrimination is reproduced for Pacific people in NZ society. In the third section, we review two current Pacific migration schemes—the Pacific Access Category scheme and the Recognised Seasonal Employer Limited Visa—and further analyse the neoliberal/capitalist exploitation these schemes entail.

---

[2] Institutional racism is understood as discrimination against minorities through unwitting beliefs, behaviours and stereotyping within a country's education, justice and health systems. Structural racism refers to wider social and political disadvantages experienced by specific social groups, for example higher rates of poverty or death (Lander 2021).

[3] NZ society and its government administration tends to categorise all Pacific migrants as one single group although limited data is available for subdivisions inside that.

From the fourth section onward, instead of analysing the (in-)feasibility of migration policies, we turn to examining governmental efforts to resolve human-induced climate problems, such as tackling rising sea levels. This section assesses both Kiribati's in-situ climate resilience building and New Zealand's efforts in helping to strengthen that resilience.

Building on these critical analyses, we conclude that climate change mitigation through labour mobility should be considered only as a last resort. Options that are likely to generate better outcomes, including a commitment on the part of developed countries to cut carbon emissions and offer assistance to strengthen the climate resilience of low-lying Pacific Island Countries (hereafter referred to as PICs), alongside the implementation of sustainable adaptation policies and practices by PICS, should be the priority. As it stands, we believe that the current New Zealand Government has failed to deliver concrete results in its Greenhouse Gas reduction, even though the country optimistically claimed to meet the Kyoto protocol's non-binding agreement.[4] At the end of this chapter, we also conclude that, as a developed country with a highly ranked national economy, New Zealand should accept a moral obligation to significantly cut its carbon emissions and engage with more effective measures.

A Brief History of Immigration and Pacific Labour in New Zealand

Like every progressive Western nation, New Zealand struggles with institutional racism and racial inequality. To understand what this means for contemporary Pacific climate migrants, we need to address New Zealand's previous approaches to migration, especially regarding Pacific labour. The

---

[4] Looking at its officially published data (New Zealand Ministry for the Environment 2021), New Zealand's overall gross emissions follow a growing trend, despite the figures for gross emissions stabilising at around 80,000 kt per year since. Recent data shows no sign of a significant drop in its gross emissions. Even if we take into consideration the offset units of carbon emissions and only speak of its net emissions, the figures still waver between 50,000 kt and 55,000 kt per year for the last eight years. And surprisingly, between 2012 and 2019, the country's net emissions are much higher when compared to the period between 2007 and 2010. (N.B. The data used in our analysis does not take into account the period after 2020, as the outbreak of the COVID pandemic, which took place in early 2020, has significantly affected the global economy and subsequently affected the rate of global greenhouse gases emissions.)

history of Pacific labour migration in New Zealand, dating from the 1960s onwards, is particularly useful to our analysis as (a) the history explains how Pacific migrants eventually became a specific "underclass" in New Zealand; and (b) it provides a contextual link to New Zealand's current migration policy and explains why "labour" migration is the New Zealand Government's preferred approach.

Early Immigration History: From European Settler Immigration (1880s) to the First Wave of Pacific Labour Migration (1960s)

From the 1880s until the mid-20th Century, New Zealand's immigration policy did not include the Pacific. Instead, it mainly emphasised the development of New Zealand as a Southern version of Britain and used a racialised immigration framework that preferentially targeted people from Britain and Ireland to populate the country (Belich 2001). This immigration approach reflected political and cultural dynamics that privileged white, English speakers of British descent (Murphy 2003). Other ethnic groups were subject to significant discrimination. For example, New Zealand's indigenous Māori population were legally British subjects after the signing of the Treaty of Waitangi in 1840 (Orange 2015) but successive New Zealand governments have created a variety of circumstances that lead to their loss of land, language, and culture, both during and after colonisation. The other significant non-European group in New Zealand prior to the 20th Century were Chinese miners (Spoonley 2015). They were invited to New Zealand to provide labour during the gold rush from 1865 to 1900 (Bradshaw 2009). However, they were often excluded from settler society and after 1882 they were also subject to a Chinese poll tax (Murphy 2003; O'Connor 2001). This discriminatory treatment of non-white ethnic groups, reflecting development of a white hegemony during the colonial period, set the stage for subsequent attitudes to immigrants in New Zealand.

The post-war period saw New Zealand begin actively recruiting migrant labour from the Pacific countries under its administration, namely Tokelau, Niue, and Rarotonga (Anae 2012). This period was characterised by labour shortages in manufacturing that immigration from Britain and the internal migration of Māori labourers could not cover (Spoonley and Bedford 2012). During the 1960s, further labour migration from Samoa, Tonga, and Fiji was initiated by employers who required workers for urban manufacturing during a time when many New Zealanders were leaving the country either permanently or for extended periods (Spoonley 2015). Pacific workers came into New Zealand as both short-term and permanent migrants (Mitchell 2003) and many became permanently settled. The demand for labour was strong up until 1974, and during this period "Pacific Islanders were generally accepted as a temporary reserve labour force" (Mitchell 2003: 158). This led many Pacific people to enter New Zealand on three-month visitor visas, obtain work illegally, and then "overstay" when their visas expired. Pacific workers were considered essential to the survival of some industries and to the function of Auckland's hospitals, so, while labour was in short supply, the authorities turned a blind eye to any illegal overstayers (Mitchell 2003). However, when the economic boom and its associated labour shortages ended, so did general acceptance of Pacific migrants in New Zealand.

Pacific Labour Migration: Dawn Raids, Backlash, and Racism Against the Pacific Migrant Community in the 1970s

The economic crisis experienced by New Zealand from 1974 onwards was triggered by the OPEC oil crisis, coupled with a drop in exports which occurred when Great Britain, New Zealand's major trading partner, joined the EEC (Nixon and Yeabsley 2010). With economic retraction and a corresponding rise in unemployment, social and cultural acceptance of Pacific immigrants in New Zealand reduced significantly and they were viewed as a threat to jobs and resources that "rightfully" belonged to New Zealanders (Anae 2012). This view reflected the centrality of "whiteness" or "European-ness" to contemporary definitions of a "real" New Zealander which were established via the racialised immigration policy pursued during the colonial period. From 1972 to 1978, letters to right-wing newspapers clearly reflect negative attitudes towards Pacific people in New Zealand, linking them to crime, housing shortages, "over-staying" (where migrants remain in the country after their visa has expired), diseases and overloading of the healthcare system. Pacific Islanders were also seen as refusing to assimilate fully into New Zealand society, and as lacking the skills to live and work in a "Western" society (Mitchell 2003).

It was not only social attitudes that changed, however. Where government agencies had previously turned a blind eye to Pacific overstayers because of the demand for labour, they now cracked down to remove them from the country. During what has become known as the "Dawn

Raid" era, police targeted Pacific people as possible visa overstayers, but did not approach those from Europe or the United States (Tokalau 2021). Targeting could occur on the street, with police asking to see a current visa and permits but was characterised by "dawn raids" where police would go to the homes of Pacific migrants very early in the morning, drag anyone suspected of being an overstayer out of bed and take them for questioning (Anae 2012). The dawn raids are now officially recognised by the New Zealand Government as an example of racial discrimination. Although the New Zealand Government has recently issued a formal apology (Tokalau 2021) this does not mitigate the disadvantages that Pacific people continue to face in New Zealand. On the contrary, institutional and structural racism against Pacific people, which has its roots in New Zealand's enduring white hegemony and all of its previous immigration policies, continues to shape the pan-Pacific community's experiences of life in New Zealand—including those who are recent migrants.

Labour-Market Immigration Frameworks and the Challenges of a Neoliberal Model Since the 1980s

Since the 1980s, New Zealand has moved away from an immigration framework that uses country of origin to target largely European migrants, to a points system[5] similar to that of Canada and Australia (Spoonley 2015). This reflects a change in emphasis to focus on attracting specific types of skilled labour rather than on including or excluding particular cultural groups. The move away from racially or geographically targeted migration might be construed as socially progressive. However, migration as an economic benefit to New Zealand, which began with the admission of Chinese miners prior to 1900 and continued with Pacific labour migration in the post-war period, remains the primary driver. The government operates a neo-liberal "light touch" model which offers little support to help migrants integrate into New Zealand and assumes labour

---

[5] New Zealand's immigration points system awards points to potential skilled migrants based on their age, possession of desirable skills and qualifications, English proficiency, and any offer of skilled work, with additional points for work offers that require residency outside of Auckland, and having a partner who is also qualified, skilled, and proficient in English. There is no consideration given to country of origin or demographic factors like race (New Zealand Immigration 2021c).

migrants will exercise self-interest to locate the best working situation (Spoonley 2015). This light touch model presents a significant number of challenges for Pacific migrants moving to New Zealand from small or subsistence economies and also for the disadvantaged Pacific communities already here. The implications of these challenges for migration as a solution to the impact of climate change in the Pacific are explored more fully below.

A Persistent Underclass? Challenges Faced by New Zealand's Pacific Community

In this section, we explore the current position of the pan-Pacific community in New Zealand, touching on the structural and institutional racism its members experience. We then demonstrate the difficulties that Pacific migrants confront when they arrive in New Zealand and become part of an existing and disadvantaged Pacific community. This analysis challenges the idea that migration is a real solution for Pacific climate change problems.

Bedford and Bedford (2010) and Thornton (2011) address migration as a tool to manage the impact of climate change in the Pacific. However, like much academic research on Pacific climate migration, they do not address a key aspect of this process, namely migrants' adaptation experiences after arriving in their new environment. Namoori-Sinclair (2020), Sin and Ormsby (2018), Thompson (2015), Gillard and Dyson (2012) and McLeod (2010) all identify a number of consistent challenges faced by Pacific migrants in New Zealand. The New Zealand Government also regularly releases statistical analyses of census and other data that show Pacific Peoples (including those born in New Zealand as well as recent migrants) have poorer social and economic outcomes than the general population. Despite this substantive documentation of challenges faced by both Pacific migrants and New Zealand's extant Pacific community, the New Zealand Government generally provides neutral or positive claims about immigration to those applying via their official channels. As a specific example, Immigration New Zealand describes the Pacific Access Category (PAC) visa scheme, offered exclusively to citizens of Kiribati, Tuvalu, Tonga or Fiji, as a way for Pacific migrants to live in New Zealand indefinitely "while also enjoying our unique lifestyle" (New Zealand Immigration 2021a). Immigration New Zealand does include some warnings that life in New Zealand may be very different from the

Pacific Islands (New Zealand Immigration 2020). However, that information is presented completely separately from the invitation to apply for entry and is not sufficient for migrants to have a full picture. We argue that the poor social outcomes faced by Pacific Peoples in New Zealand, coupled with New Zealand's current approach to immigration contribute to a number of ongoing challenges for Pacific migrants, including people from Kiribati. This makes climate migration to New Zealand at best a difficult choice for Pacific citizens.

The Pacific Community in New Zealand

For statistical and some administrative purposes, New Zealand treats all people with Pacific heritage, including those from Kiribati, as a single ethnic group, even though this group encompasses multiple, related Pacific cultures and countries of origin (Matika et al. 2021). In the most recent 2018 census, the term "Pacific Peoples" was used to refer to anyone with Pacific heritage (Stats NZ, n.d.) This group is currently the fourth-largest ethnic grouping at 8.1%.[6] It is also recognised culturally, with the ethnicity "Pacific Islander" or "Pasifika" applied to both recent immigrants and New Zealand citizens of Pacific origin, regardless of which part of the Pacific they or their family come from.

Institutional Racism Experienced by Pacific People in New Zealand

Recent census data analysis indicates that, as a group, Pacific Peoples—including recent migrants—are significantly disadvantaged compared to the general population. They experience worse outcomes across measures such as education, income, and job category. They are proportionally under-represented in professional and managerial work and over-represented in manual labour and care work (Stats NZ, n.d.). Their median income is 25% lower, and they are 50% less likely to have a bachelor's degree than the general population (Stats NZ, n.d.). Pacific people also have significantly worse health outcomes, associated with

---

[6] For readers unfamiliar with the cultural make-up of New Zealand, Pacific people form the fourth-largest cultural grouping, making up 8.1% of the population in 2018. Pakeha (white New Zealanders) are the most numerous at 70.2%, Maori (indigenous New Zealanders) are next at 16.5% and Asian New Zealanders follow at 15.1%. All other ethnicities together form just 2.7% of the total population (Stats NZ 2019).

uneven access to healthcare and lifestyles correlated with poverty and low income (Came et al. 2019). For example, they are significantly more likely to smoke cigarettes and to be obese than New Zealanders of European heritage (Marriott and Alinaghi 2021). These statistically worse outcomes, demonstrate ongoing institutional racism in New Zealand which extends to anyone of Pacific heritage. I-Kiribati migrants to New Zealand, therefore, arrive into a social and political context which impedes their long-term likelihood of achieving a decent standard of living. They also face immediate practical challenges when establishing themselves in their new home, and these additionally impede their long-term chances of success.

I-Kiribati Experiences of Life in New Zealand

Confluent with New Zealand's "light touch" approach to immigration discussed above, I-Kiribati migrants receive no specific assistance to help them transition to life in New Zealand's complex economy. This is despite coming from a largely subsistence economy in which only 20% of the population are formally employed (Pretes and Gibson 2008).[7] They must also accustom themselves to a neoliberal environment which emphasises payment for services and contrasts strongly with the communal, local and community-centred *maneaba* system[8] of social support and governance they are accustomed to (Namoori-Sinclair 2020; Sofield 2002). Many I-Kiribati experience "shock and surprise" when they first arrive in New Zealand as new migrants, particularly if they have not previously travelled outside of Kiribati. Some I-Kiribati migrants report that they expected a free house and employment to kick-start their settlement (Thompson 2015: 131). Many of Thompson's interview participants also expressed surprise at the importance of money in the New Zealand economy. In Kiribati, economic transactions are a mix of "money, barter, self-dependence or extension of gifts" (Thompson 2015: 132), and moving

---

[7] While 20% of Kiribati's population are in formal employment, the other 80% combine subsistence (farming) activity with financial support from family, both local and overseas.

[8] In Kiribati the *maneaba* is a central meeting house which is the locus of village life and the basis of governance at both island and national level. Regular village meetings ensure that, in an atoll environment with limited resources, all members of the community have their needs met. Decisions are made, and authority applied, by consensus. The *maneaba* is based on the political concept that the community must be collectively responsible for its own management.

into an environment which requires money for all transactions is a significant adjustment, particularly when new migrants are usually unable to bring much cash from Kiribati's largely informal economy, and may not yet be formally employed (with access to wages or salary) in New Zealand.

In the absence of anticipated formal support from government agencies, most new migrants rely on assistance from social networks of I-Kiribati already settled in New Zealand (McLeod 2010; Thompson 2015), as well as Pacific churches and other community organisations (McLeod 2010). This takes the form of accommodation, assistance to gain employment, and social support. It may also include assistance with unexpected financial demands, such as high electric bills or specialist medical costs (Thompson 2015). Providing support like this is common among many Pacific cultures in New Zealand (Gillard and Dyson 2012).

Accommodation is a key priority and provision of accommodation is recognised as one of the most vital forms of support by new I-Kiribati migrants (Thompson 2015: 141). New migrants in New Zealand must navigate a nationwide property market characterised by a vicious housing shortage and high-cost accommodation. Real-estate in Auckland, the largest city, is estimated as the fourth least affordable in the world (Cox 2021). However, even before the current housing crisis, social networks of I-Kiribati in New Zealand have provided accommodation to new arrivals. This might include arranging separate housing, or helping migrants save money to get into their own home. Very commonly it involves hosts providing a room in their house to a family while they get on their feet—often for months at a time (Thompson 2015; McLeod 2010). In fact, it is not unusual for a host family to have multiple families staying, one per bedroom (Gillard and Dyson 2012). For I-Kiribati and Pacific people generally, there is a strong social and cultural obligation to provide support, and multiple-family homes are considered normal (Gillard and Dyson 2012). However, several families living in one home is considered over-crowding by the rest of New Zealand and may create problems with private landlords or with the government's social housing property managers.[9] It can also be challenging for the host family who are not always in a stable financial position themselves. The additional household costs create pressure on host adults to work multiple jobs and sometimes for older children in host families to leave school and go

---

[9] This in turn disincentivises landlords from renting to Pacific families and makes finding permanent accommodation more difficult for this group.

into paid work rather than completing their education (McLeod 2010). Hosting is considered culturally appropriate by I-Kiribati and is usually done very willingly. However, it is also a practical necessity given the lack of government resources provided to new migrants, and it places a significant and continuing social and financial burden on the New Zealand-based I-Kiribati community.

As discussed above, I-Kiribati migrants arrive from a largely subsistence economy and need to adapt and find work in New Zealand's complex economic environment. Many find the adjustment to formal, paid labour challenging. There can be poor or incomplete communication about what to expect from working conditions in New Zealand (Thompson 2015). In Kiribati, "the atmosphere is very relaxing, and people could spend hours chatting without worrying about their bills or rents" (Korauaba 2011: 19). While I-Kiribati social networks do assist new migrants to find work (Thompson 2015) it can be difficult for them to get skilled, well-paid work in New Zealand. If I-Kiribati have qualifications, most are not recognised, and accessing tertiary education is expensive (Gillard and Dyson 2012). In addition, many I-Kiribati do not speak English with complete fluency. Thompson (2015) found that none of the participants who could have benefitted from accessing language training had done so at the time of her interviews. There is a strong correlation between proficiency in English and the earning power of new migrants in New Zealand (The Office of Ethnic Affairs 2013). Low English language proficiency makes it difficult to get work at all and it can result in migrants becoming stuck in a catch-22 where their low-paid work (like night-time cleaning) does not provide a chance to develop their English skills, thus keeping them from accessing better employment opportunities. Poor English also creates a strong barrier to entry into tertiary training for I-Kiribati migrants (Thompson 2015). This inability to access professional training is a further and significant barrier to well-paid employment.

The challenges faced by I-Kiribati to find adequate accommodation and skilled work in New Zealand demonstrate how easy it is for them to become part of a Pacific underclass, experiencing substandard accommodation and inter-generational poverty linked to under-employment. Given the negative aspects of relocating to New Zealand from the Pacific, it makes sense to further assess its current migration and resettlement schemes as well as to consider whether an in-situ resilience and adaptation plan can offer a real alternative for those who prefer not to relocate.

ADDRESSING THE EFFICACY OF PACIFIC MIGRATION AND RESETTLEMENT SCHEMES

Politically speaking, global climate change problems are often framed with, and driven by, humanitarian discourses. On the issue of climate displacement, the generally accepted ethical way to resolve this issue is to encourage developed countries to deliver climate justice by offering humanitarian visas for people migrating from low-lying atoll nations. Since 2010, climate migration has become a widely promoted political strategy in delivering climate justice, and in 2016, the Mary Robinson Foundation further published its position paper to reinstate the necessity of adopting such an approach (Mary Robinson Foundation – Climate Justice 2016). However, on a theoretical level, this climate migration approach has been insufficiently challenged.[10] Moreover, in practice, the delivered result hardly reflects the true meaning of humanitarianism. In this section, we review New Zealand's current offer of migration schemes for low-lying Pacific islanders, as this can help us better understand the country's level of commitment to delivering climate justice.

Although New Zealand prides itself on being at the forefront of a quick humanitarian response to the call of climate "refugees" or migrants (Dempster and Ober 2020), one may be surprised to discover that only a very limited number of places for long-term resettlement are offered and they remain extremely difficult to access. We argue that, overall, NZ's current migration schemes for Pacific climate migrants ultimately reflect its underlying economic approach to immigration rather than addressing climate change in a meaningful way. Today, the NZ government offers two types of visas exclusive to Pacific migrants: the Pacific Access Category scheme (hereafter referred to as PAC) and the Recognised Seasonal Employer Limited Visa (hereafter referred to as RSE). The NZ government claims that both schemes have directly or indirectly helped address

---

[10] Chris Methmann and Angela Oels point out that recent discourses treat "climate-induced migration" as a rational strategy to adapt to unavoidable levels of climate change, such that the relocation of millions of people is considered acceptable. However, their work further criticises how this climate migration strategy is largely promoted by liberal governments in highly developed countries, which ultimately exercise their sovereign power and regulation through liberal biopower (Methmann and Oels 2015).

climate displacement problems in the Pacific region. However, we argue that neither of these schemes can be considered an appropriate solution.[11]

Pacific Access Category Scheme (PAC)

A benign intention to promote human welfare may be recognised as underlying the NZ government's setting up of the Pacific Access Category (PAC), as it guarantees an indefinite length of stay in New Zealand and allows visa holders to both work and study in New Zealand. The conditions of entry also cover one's partner and dependent children under the age of 24, which shows that a long-term family integration of those likely to become part of the workforce in NZ society is being targeted (New Zealand Immigration 2021a). The NZ government and some scholarly research praise the scheme as a positive and forward-thinking strategy for dealing with Pacific climate displacement (Higuchi 2019). However, the scheme ultimately fails as a humanitarian endeavour. The PAC scheme operates by a completely random selection, via a lottery ballot system. But while this may appear to embody equality and fairness for all applicants, the selection mechanism can also be criticised for overlooking those who are in urgent need of relocation. Moreover, the maximum annual intake of Pacific migrants is 625 people in total, among whom *only 75 I-Kiribati citizens per year* have the possibility of being selected (New Zealand Immigration 2021a). For a country like Kiribati that faces the critical threat of climate change and has a strong claim to climate justice, the 75 places offered seem completely insufficient to satisfy the need. Moreover, though qualified as a "humanitarian" approach, the NZ governmental website opts to put forward strong messages about how the New Zealand work ethic should be respected by these in-bound Pacific migrants. On the website, a brief summary explicitly states, "[b]e prepared to work hard in New Zealand. It is important to be responsible and committed to your job – it will help you settle well in New Zealand. Do not waste this opportunity" (New Zealand Immigration 2020). To a certain extent,

---

[11] We would also like to note that, as a result of the COVID pandemic, the NZ government has suspended the PAC scheme since early 2020 and imposed a number of restrictions on the RSE visa for Pacific applicants. Although these precautionary measures are perhaps necessary and have occurred in an exceptional time, the current situation also raises questions regarding the sustainability of the climate migration route as a long-term solution.

the logic implied by these statements is that offering the PAC scheme to Pacific islanders is a useful way for New Zealand to target the low-cost labourers its economy needs. Last but not least, the expensive fees associated with this visa application can also be interpreted as a way to dissuade applications for long-term resettlement. In stage 1 of the PAC application, one is required to register for the ballot, which costs AUS$ 94.60 (equivalent to approximately US$71). If one makes it to stage 2 of the application process, FJ$1907 (equivalent to US$918) will need to be paid to complete the application process (New Zealand Immigration, n.d.[a]). For a country like Kiribati that has very low average Gross National Income (GNI), these costly fees make it almost impossible for most I-Kiribati families to apply.[12] It is therefore not hard to see that although the PAC scheme allows for an intake of 75 I-Kiribati migrants annually, the take-up rate of this scheme has generally been far below this number, indicating its failure to meet the needs of the people it claims to serve (Wyett 2013).

Recognised Seasonal Employer Scheme (RSE)

Many would argue that apart from the PAC scheme, the NZ government also offers the Recognised Seasonal Employer Scheme (RSE) as an alternative route to help Pacific migrants and their family, and that this scheme can be indirectly beneficial to those who are subjected to climate displacement. Although the RSE does not offer a permanent resettlement opportunity in New Zealand, it allows the Pacific migrant workers in New Zealand to send remittances back home. This could, theoretically speaking, strengthen resilience in countries like Kiribati or Tuvalu which are prone to rising sea levels and coastal erosion. On the

---

[12] According to *UNDP's Human Development Report*, Kiribati's Gross National Income (GNI) per capita on average is $4,260 (2019 census) (UNDP Human Development Report 2021a). To put this figure into perspective, it is ten times lower than New Zealand's GNI per capita, which is $40,799 (2019 census) (UNDP Human Development Report 2021b). *The 2006 Kiribati Household Income and Expenditure Survey* revealed relatively small contrasts, among the various groups of islands making up the country, in terms of annual per capita income, ranging from A$1,053 in the Southern Gilbert group to A$1,531 in Southern Tarawa where the capital is located. The survey also revealed that households in Kiribati spend more than they earn (are indebted), and that a large majority of the islanders are financially supported by the small minority with a regular income (United Nations Committee for Development Policy 2018).

NZ Foreign Affairs and Trade website, the government cheers the "success" of the RSE scheme: "In 2018/19, 11,168 Pacific workers came to New Zealand under the RSE. Pacific seasonal workers in New Zealand remit over NZ$41million [c. US$ 28 million] in remittances a year" (New Zealand Ministry of Foreign Affairs and Trade 2021). However, one must acknowledge that this scheme is likely to contribute very little help to individuals who require assistance for climate relocation or on-site resilience building. As Lacey Allgood and Karen E. McNamara (2017) demonstrate, "reliance on overseas remittances – [only] 27 per cent [of] respondents – was the least utilized method by households to respond to the impacts of a changing climate" (378). Asking for government assistance, reducing expenses, or working in a second job are all more frequently employed by I-Kiribati households. Allgood and McNamara express surprise over this finding, as it contradicts previous research which generally highlights the importance of overseas remittances and claims that remittances could significantly help improve the day-to-day lives of many I-Kiribati families (378; Browne and Mineshima 2007; Borovnik 2006). Allgood and McNamara's research thus points out the misconceptions surrounding overseas remittances and their limited contribution to mitigating climate impacts in PICs like Kiribati.

Besides this unexpected finding, one can also criticise the RSE scheme for a number of other reasons. First of all, the scheme offers only an extremely brief period of stay, up to only 7 months in any 11-month period for citizens of certain Pacific nations (New Zealand Immigration 2021b). Despite Kiribati and Tuvalu citizens having the privilege of an extra 2 months, the short duration of the stay would hardly enable any labour migrants to explore or plan ahead for any resettling possibilities. The government website for RSE applications clearly states that one must leave New Zealand before the visa expires, and one *cannot* appeal to the Immigration tribunal to stay in New Zealand. Secondly, the seasonal working opportunities are only provided in the horticulture and viticulture industries, and neither of these industries would be able to offer significant income for its employees. With these factors in mind, the RSE scheme is specifically designed to "extract" surplus labour value, and thus can be regarded as a form of exploitation of low-cost human labourers from PICs. Finally, the RSE scheme does not include partners or dependent children in the same application. Compared to the PAC scheme, this rule explicitly excludes the aspect of "humanitarian aid". One may thus rightfully argue that this visa scheme targets only individual human labour and cannot be qualified as belonging to a humanitarian approach to Pacific migrants.

Summarising our analysis of current migration pathways New Zealand offers to Pacific islanders, we consider that neither the PAC nor RSE schemes are able to effectively deliver positive outcomes for Pacific climate migration. The PAC scheme has a variety of limitations and the RSE scheme ultimately creates a mechanism that perpetuates the exploitation of low-cost human labour. It is also interesting to note that, in the last decade, there has been an increasing intake of Pacific migrants via the RSE scheme, from 8,000 offered places in 2009 to 14,400 places in 2020/2021 (New Zealand Immigration, n.d.[b]). In contrast, the places offered to I-Kiribati and Tuvaluans through the PAC scheme remain unchanged since it was introduced in 2002. The significant increase in places available to migrants entering New Zealand through the RSE scheme shows that its main interest still lies in exploiting migrant labour, rather than providing humanitarian aid for climate relocation.

CRITICAL ASSESSMENT OF KIRIBATI'S IN-SITU ADAPTATION APPROACH

The preceding analysis looked at Pacific migration in New Zealand, in both its historical and contemporary contexts. We outlined the case of Kiribati to demonstrate that the migration pathways offered do not reflect a genuine effort on the part of New Zealand to assist low-lying atoll countries to resolve issues of climate displacement. Instead, they reflect a neoliberal approach to cheap human-labour extraction. In the following section, we analyse policies of on-site resilience-building in Kiribati, which is critical if we are to understand how the discourse of "climate emergency" can be shaped and re-shaped in vulnerable Pacific islands. Through these discursive changes, one recognises that claims of "climate change in the Pacific" can also be a form of political leverage to gain more access to international funding. Reviewing different policies that have been put forward by the Kiribati government in the last decade, we argue that neither the migration route nor the current in-situ adaptation approach is up to the task of mitigating severe impacts of climate change in Kiribati (e.g., rising sea levels, seawater inundation, fresh water salination, food crop scarcity, coastal erosion, etc.).

Kiribati's Climate Adaptation Discourse: Environmental or Economic Sustainability?

Under Anote Tong's presidency (2003–2016), the main policy used to mitigate the impacts of climate change in Kiribati was "migration with dignity". From 2017 onward, with Taneti Maamau assuming the presidency, Tong's "migration with dignity" policy was immediately aborted. President Maamau made clear in his public address at the COP23 UN Climate Change Conference that the new government will dedicate its efforts solely to adaptation strategy.

'I want to make it very clear at the outset that my Government has decided to put aside the misleading and pessimistic scenario of a sinking/deserted nation, and has replaced it with a bold scenario filled with great faith in the Mighty Hand, that made our islands, coupled with our people's unwavering love for their home land, and great determination to fight and/or adapt to climate change' (Maamau 2017).

In changing tack in this manner, one may argue that Maamau and his government are embracing a more holistic and positive approach, since his statements outline the importance of reconnecting the I-Kiribati people to their beloved homeland, and further give priority to reenforcing the country's environmental wellbeing and resilience in the face of climate change. Indeed, it could be argued that this approach aims for the long-term development of a healthy and climate-resilient environment. However, when one investigates Maamau's public address more closely, one understands that this "adaptation" or "in-situ resilience building" vision does not aim to seek mitigating solutions to climate change. In the same document, Maamau (2017) further affirms:

'To this end, my Government is calling on this Conference, to recognize the significance of this new policy as a platform for the Kiribati Government's ambitious 20 year plan or KV20, to build and develop the nation in the face of climate change – *that focuses on harnessing our resources on fisheries and tourism.* The continued conversation and predictions for Kiribati to sink in future are not only de-empowering but also contradictory to our current efforts to build our islands and transform the lives of our people into a resilient, wealthy, healthy, and secured nation in line with our KV20.' (emphasis mine).

Maamau's statement harshly criticises the former government's belief in a "sinking island" and its political investment in the "migration with dignity" policy. He even employs the word, "de-empowering", to emphasise his disagreement. However, apart from this criticism, not much emphasis has been placed on the new government's dedication to tackle the climate problem. For instance, Maamau's statement does not address the core issue of rising global temperatures at all and nor does it ask for other developed countries to make a joint effort to cut GHG emissions. When Maamau highlights the need to "transform the lives of our people into a resilient, wealthy, healthy and secured nation", the wording implies that the current government is more enthusiastic about building a "developed" country through making economically productive use of its own resources, rather than achieving "environmental sustainability". Researchers have in the past criticised that, since the publication of the Brundtland Report,[13] the key concept of "sustainable development" has often been instrumentalised as a "new jargon phrase in the development business" (Adams 2009: 6; Conroy 1988: xi). The same logic applies to the current Kiribati government's deliberate adoption of this rather vague and rhetorical framing of terms such as "sustainable development" or "climate resilience". As a matter of fact, Maamau's statement focusses on Kiribati's economic need to prioritise the development of fisheries and tourism resources, which lie quite a way from tackling the problems of climate change. Ultimately, this objective can only be interpreted as an address to "economic sustainability" but not "climate or environmental sustainability".

KJIP: Rephrasing Climate Change Issues into Risk Management Issues

In 2019, two years after Maamau's official address at the UN Climate Conference, the Kiribati government finally published the *Kiribati Joint Implementation Plan for Climate Change and Disaster Risk Management* (hereafter referred to as *KJIP*). Though this document highlights that climate change mitigation initiatives are a national priority, no *actual* adaptation actions were achieved between 2017 and 2019. As the same

---

[13] *The Brundtland Report*, also known as *Our Common Future*, is a document published in 1987 by the World Commission on Environment and Development (WCED). The report introduces the concept of "sustainable development" and sets out a global agenda for change. This has created a major shift in culture and policy. In the *Brundtland Report*, "sustainable development" is seen as "development that meets the needs of the present without compromising the ability of future generations to meet their own needs".

document shows, the estimated timeframe for implementing, monitoring, and evaluating such actions would only begin after 2019, because the government requires *two full years* to revisit the Tong government's created policies and ensure that forward planning is in alignment with the Kiribati Vision 20 (KV20) (Government of Kiribati 2019).

There are twelve climate mitigation strategies listed in the document, and most are coupled with at least one or two key national adaptation priorities. Undeniably, there are some positive features of this implementation plan. For instance, an investment in "greening" certain industries (i.e., implementing greening initiatives in small or medium-sized private enterprises in tourism, trade, transport, imports, and exports) is suggested for integration into all future planning (Government of Kiribati 2019). Nonetheless, reviewing all strategies and the adaptation priorities, the main message conveyed in this document is that the government generally focusses more on security issues and disaster risk management than on climate change and environmental sustainability. The *KJIP* carefully rewords climate change issues into risk management issues. On paper, the government may appear to emphasise its concern for long-term solutions to climate change, but in reality, it mainly targets short-term responses to sudden large-scale climate events.

One specific example that can illustrate our argument relates to the current government's investment in setting up a new meteorological centre. In line with the drafted action plan (Government of Kiribati 2019), the government sees professional training and improvement of facilities for environmental data collection or analysis as essential, because they deliver early warnings for many climate-related hazards (e.g., hurricanes, king tides, soil erosion, etc.), and provide regular reports on the environmental outlook. The plan further emphasises the need to "strengthen the capacity to collect, assess and analyse relevant *agro-meteorological data* and [the] impacts on crop yields" (Government of Kiribati 2019: 110, emphasis mine). Nevertheless, cross-referencing with the *Kiribati Meteorological Service Strategic Plan & Framework* (Kiribati Meteorological Service 2021), a published national action plan on the website of the Global Framework of Climate Service, one may be intrigued to learn that the top two key achievements of the period 2017–2019 that are praised by the government are the installation of a new meteorological station at the Cassidy International Airport (Kirimati/Christmas Island) and the upgrade of meteorological equipment at both Bonriki (Tarawa) airport and Cassidy International Airport. If the government is outlining large-scale climate risk prevention, then why would it prioritise working to improve meteorological facilities at the airports, rather than other more urgently needed areas? This raises the doubt that the current government has always intended to give priority to developing its tourism industry over genuine climate mitigation and adaptation actions. Furthermore, one may also wonder how the method of risk management through early-warning systems can contribute to resolving global climate change. Ultimately, this method can only inform people of how and when to take precautionary measures, but it can never reduce the frequency of large-scale climate disasters.

In sum, Kiribati's current adaptation strategy exhibits limited engagement with resolving human-induced climate problems. We argue that Kiribati's current policymaking focus on in-situ resilience serves mainly as a "tick-box" political gesture to fit in with global climate change discourse. We want to emphasise that both the New Zealand and Kiribati governments are placing economic benefits over environmental outcomes in their policy approaches. The former has mainly focussed on human mobility and extraction of the surplus value of migrant labourers, and the latter replaces the discourse of climate sustainability with economic sustainability. Ultimately, there is a significant gap between the discourse of climate change and environmental sustainability and its application in the policymaking of the region.

CONCLUSION

At the beginning of this chapter, we raised the critical question of whether a "Western" climate justice framework could deliver meaningful responses to climate displacement issues experienced by the vulnerable island states in the Pacific. On this very topic, we also asked: has a climate migration approach—the prioritised solution in a climate justice framework—become problematic when applied in the South Pacific context? Through our findings and analyses, we argue that, to a large extent, the approach of climate justice proposed by civil societies or scholars mainly relies on human-rights-based values, but this does not work when it is applied in the South Pacific. Climate justice could potentially be praised as a good framework, and climate migration as a feasible solution, when the social and economic environments of the country of emigration and the country of destination are relatively similar. And it may also work positively when there is little or no history of a hierarchical power relationship between the two countries. But this is simply not the reality for the vulnerable PICs we examined. Using the case of Kiribati, we see further aggravation of social discrimination and the widening of economic inequality as part of the consequences of this climate migration approach.

Our argument here does not aim to suggest that human rights-based values should be entirely disregarded or ignored when articulating answers to the climate displacement challenge in the Pacific. However, we question this approach because we consider the development of a climate justice framework follows a logic that only leads to a short-sighted, "anthropocentric" responseto the Pacific climate crisis,[14] which also exacerbates existing power relations between the "West" and its "periphery". The inherent logic of climate justice steers away from eco- or bio-centric positions to address the current climate challenge (Katz and Oechsli 1993; Chakrabarty 2021).[15] And ultimately, it can only put forward an insufficient solution for those islands which are already suffering extreme environmental challenges.

To summarise this chapter, we have examined the history of New Zealand's immigration policies, showing a persistent pattern of institutional and structural racism that negatively impacts the Pacific community living in New Zealand today. New Zealand's current neoliberal "light-touch" immigration approach reflects its overall economic approach to immigration policy and means new Pacific migrants, including those from I-Kiribati, receive little government support to adapt to life in a complex economy, ensuring they quickly become part of a socio-economically disadvantaged Pacific community after they arrive. This suggests that escaping the impacts of climate change in Kiribati by migrating to New Zealand is simply exchanging one set of problems for another. We further argue that the existing mechanisms enabling Pacific migration into New Zealand (i.e., working and migration visas) have the effect (whether

---

[14] As Mary Robinson, the founder of the Mary Robinson Foundation- Climate Justice, firmly states in an interview, "Climate change is often framed as an environment issue... [and] primarily a technical issue, a scientific issue. Climate justice, with its foundations in human rights and development, takes a different approach. Climate justice makes climate change an issue about people" (Gearty 2014).

[15] Dipesh Chakrabarty, in his recent published book, highlights that it is inadequate to only seek answers to the problems of justice between humans and focus on human welfare (i.e., a strictly anthropocentric justice) if we attempt to resolve the anthropogenic climate change in our time (Chakrabarty 2021).

intended or not) of transforming climate migrants into economically exploitable labour migrants. As this is not a desirable outcome, we turn to an examination of Kiribati's adaptation policies, considering whether they can support sustainable adaptation to the impacts of climate change and reduce the need for migration. Our analysis concludes that the Kiribati government's current adaptation plan is focused on local economic development and mitigation of short-term disasters, rather than meaningful long-term adaptation that will enable the population to stay safely in Kiribati.

With this in mind, we suggest that if the current New Zealand government is genuinely interested in tackling the problems associated with climate change in the Pacific, then it should: (1) prioritise emissions reduction to meet the limited target of 1.5 degrees of global average temperature rise set out in the Paris Agreement; (2) increase assistance to low-lying Pacific Island Countries, including Kiribati, to build or reinforce resilience against foreseeable environmental and climate risks and avoid the last resort of Pacific islanders being "displaced" or "forced to migrate" to another country); (3) and as a last resort only, offer better supporting systems for Pacific migrants (including both climate and labour migrants) to transition well to life in New Zealand. Likewise, we recommend that the Kiribati government reconsider its political position with regard to global climate change's impacts on small vulnerable island states and further, set up effective communication channels with local inhabitants to foster an open conversation on the subject. Moreover, instead of reframing Kiribati's current climate challenges as security or disaster risk management issues and prioritising economic sustainability, the government should urgently revisit its policy approaches and give more weight to long-term planning for environmental sustainability.

REFERENCES

Adams, W.M., 2009. *Green Development*. London and New York: Routledge, pp. 1–25.

Allgood, L. and McNamara, K., 2017. Climate-Induced Migration: Exploring Local Perspectives in Kiribati. *Singapore Journal of Tropical Geography*, 38(3), pp. 370–385.

Anae, M., 2012. All Power to the People: Overstayers, Dawn Raids and the Polynesian Panthers. In: S. Mallon, K. Māhina-Tuai and D. Salesa, ed., *Tangata o le Moana: New Zealand and the People of the Pacific.* Wellington: Te Papa Press, pp. 221–239.

Bedford, R. and Bedford, C., 2010. International Migration and Climate Change: A Post-Copenhagen Perspective on Options for Kiribati and Tuvalu. In: B. Burson, ed., *Climate Change and Migration South Pacific Perspectives.* Wellington: Institute of Policy Studies School of Government Victoria University of Wellington, pp. 89–134.

Belich, J., 2001. *Paradise Reforged: A History of the New Zealanders from the 1880s to 2000.* London: Penguin Books.

Borovnik, M., 2006. Working Overseas: Seafarers' Remittances and Their Distribution in Kiribati. *Asia Pacific Viewpoint*, 47(1), pp. 151–161.

Bradshaw, J., 2009. *Golden Prospects: Chinese on the West Coast of New Zealand.* Greymouth: Shantytown.

Browne, C. and Mineshima, A., 2007. Remittances in the Pacific Region. *International Monetary Fund Working Paper*, 07/35, IMF, Washington DC. [Online] Available at: https://www.imf.org/external/pubs/ft/wp/2007/wp0735.pdf [Accessed 7 July 2021].

Came, H., McCreanor, T., Haenga-Collins, M. and Cornes, R., 2019. Māori and Pasifika Leaders' Experiences of Government Health Advisory Groups in New Zealand. *Kōtuitui: New Zealand Journal of Social Sciences Online*, 14(1), pp. 126–135. Available at: https://www.tandfonline.com/doi/full/10.1080/1177083X.2018.1561477 [Accessed 20 July 2021].

Chakrabarty, D., 2021. *The Climate of History in a Planetary Age.* Chicago: The University of Chicago Press, p. 48.

Conroy, C., 1988. Introduction. In: C. Conroy and M. Litvinoff, ed., *The Greening of Aid: Sustainable Livelihoods in Practice.* London: Earthscan, pp. xi–xiv.

Cox, W., 2021. *Demographia International Housing Affordability 2021 Edition.* Urban Reform Institute and The Frontier Centre for Public Policy, pp. 1–16. [Online] Available at: https://urbanreforminstitute.org/wp-content/uploads/2021/02/Demographia-International-Housing-Affordability-2021.pdf [Accessed 2 September 2021].

Dempster, H. and Ober, K., 2020. *New Zealand's "Climate Refugee" Visas: Lessons for the Rest of the World.* DEVPOLICY BLOG. [Online] Available at: https://devpolicy.org/new-zealands-climate-refugee-visas-lessons-for-the-rest-of-the-world-20200131/ [Accessed 19 June 2020].

Gearty, C., 2014. An Interview with Mary Robinson, President of the Mary Robinson Foundation – Climate Justice. *Journal of Human Rights and the Environment*, 5(0), pp. 18–21. [Online] Available at: https://www.elgaronline.com/view/journals/jhre/5-0/jhre.2014.02.03.xml [Accessed 5 April 2021].

Gillard, M. and Dyson, L., 2012. *Kiribati Migration to New Zealand: Experience, Needs and Aspirations.* Wellington: Impact Research/Presbyterian Church of Aotearoa New Zealand.

Government of Kiribati, 2019. *Kiribati Joint Implementation Plan for Climate Change and Disaster Risk Management (KJIP): 2019–2028.* Government of Kiribati, pp. 67, 70, 110.

Higuchi, E., 2019. Climate Change Policies and Migration Issues of New Zealand and Australia. *OPRI Perspectives,* (2), pp. 1–6. [Online] Available at: https://www.spf.org/en/global-data/opri/perspectives/prsp_002_2019_higuchi_en.pdf [Accessed 13 July 2021].

Katz, E. and Oechsli, L., 1993. Moving beyond Anthropocentrism. *Environmental Ethics,* 15(1), pp. 49–59. [Online] Available at: https://www.pdcnet.org/enviroethics/content/enviroethics_1993_0015_0001_0049_0059 [Accessed 6 April 2021].

Kiribati Meteorological Service, 2021. *Kiribati Meteorological Service Strategic Plan & Framework.* Global Framework for Climate Service. [Online] Available at: https://gfcs.wmo.int/sites/default/files/Kiribati_NSP.pdf [Accessed 13 May 2021].

Korauaba, T., 2011. *Report from a Kiribati Households Survey in New Zealand 2009/2010.* Auckland: Pacific Micronesian Foundation.

Lander, V., 2021. *Structural Racism: What It Is and How It Works.* The Conversation. [Online] Available at: https://theconversation.com/structural-racism-what-it-is-and-how-it-works-158822 [Accessed 6 August 2021].

Maamau, T., 2017. *His Excellency Beretitenti Taneti Maamau's Statement.* 15 November, COP 23, Bonn. [Online] Available at: https://unfccc.int/sites/default/files/kiribati_cop23cmp13cma1-2_hls.pdf [Accessed 02 June 2021].

Marriott, L. and Alinaghi, N., 2021. Closing the Gaps: An Update on Indicators of Inequality for Māori and Pacific People. *The Journal of New Zealand Studies,* (NS32). [Online] Available at: https://doi.org/10.26686/jnzs.iNS32.6863 [Accessed 5 September 2021].

Mary Robinson Foundation – Climate Justice, 2016. *Protecting the Rights of Climate Displaced People.* [Online] Available at: https://www.mrfcj.org/wp-content/uploads/2016/07/Protecting-the-Rights-of-Climate-Displaced-People-Position-Paper.pdf [Accessed 14 May 2021].

Matika, C., Manuela, S., Houkamau, C. and Sibley, C., 2021. Māori and Pasifika Language, Identity, and Wellbeing in Aotearoa New Zealand. *Kōtuitui: New Zealand Journal of Social Sciences Online,* 16(2), pp. 396–418.

McLeod, D., 2010. Potential Impacts of Climate Change Migration on Pacific Families Living in New Zealand. In: B. Burson, ed., *Climate Change and Migration South Pacific Perspectives.* Wellington: Institute of Policy Studies School of Government Victoria University of Wellington, pp. 135–158.

Methmann, C. and Oels, A., 2015. From 'Fearing' to 'Empowering' Climate Refugees: Governing Climate-induced Migration in the Name of Resilience. *Security Dialogue*, 46(1), pp. 51–52, 63–64.

Mitchell, J., 2003. *Immigration and National Identity in 1970s New Zealand*. Ph.D. The University of Otago.

Murphy, N., 2003. Joe Lum v. The Attorney General: The Politics of Exclusion. In: M. Ip, ed., *Unfolding History, Evolving Identity: The Chinese in New Zealand*. Auckland: Auckland University Press, pp. 48–68.

Namoori-Sinclair, R., 2020. *The Impact of PAC Policy on Pacific Women's Health and Wellbeing: The Experiences of Kiribati Migrants*. Ph.D. Victoria University of Wellington.

New Zealand Immigration. 2021a. *Information About: Pacific Access Category Resident Visa*. [Online] Available at: https://www.immigration.govt.nz/new-zealand-visas/apply-for-a-visa/about-visa/pacific-access-category-resident-visa [Accessed 19 July 2021].

New Zealand Immigration. 2021b. *Information About: Recognised Seasonal Employer Limited Visa*. [Online] Available at: https://www.immigration.govt.nz/new-zealand-visas/apply-for-a-visa/about-visa/recognised-seasonal-employer-limited-visa [Accessed 19 July 2021].

New Zealand Immigration. 2021c. *Points Indicator for Skilled Migrant Expression of Interest*. [Online] Available at: https://www.immigration.govt.nz/new-zealand-visas/apply-for-a-visa/tools-and-information/tools/points-indicator-smc-28aug [Accessed 3 September 2021].

New Zealand Immigration, 2020. *Moving to New Zealand with a Pacific Access Category (PAC) Resident visa*. [Online] Available at: https://www.newzealandnow.govt.nz/choose-new-zealand/compare-new-zealand/moving-from-the-pacific-islands/moving-with-a-pacific-access [Accessed 3 September 2020].

New Zealand Immigration. n.d.[a]. *Fees, Decision Times and Where to Apply*. [Online] Available at: https://www.immigration.govt.nz/new-zealand-visas/apply-for-a-visa/tools-and-information/tools/office-and-fees-finder [Accessed 6 August 2021].

New Zealand Immigration. n.d.[b]. *Recognised Seasonal Employer (RSE) Scheme Research*. [Online] Available at: https://www.immigration.govt.nz/about-us/research-and-statistics/research-reports/recognised-seasonal-employer-rse-scheme [Accessed 20 June 2021].

New Zealand Ministry for the Environment, 2021. *NZ's Interactive Emissions Tracker*. [Online] Available at: https://emissionstracker.mfe.govt.nz/#NrAMBoE4BYF12ARnAIgHIFMAuL7AEzj6iIBseAHKrrEA [Accessed 15 July 2021].

New Zealand Ministry of Foreign Affairs and Trade, 2021. *Labour Mobility*. [Online] Available at: https://www.mfat.govt.nz/en/trade/free-trade-agreements/free-trade-agreements-in-force/pacer-plus/labour-mobility/ [Accessed 19 July 2021].

Nixon, C. and Yeabsley, J., 2010. *Overseas Trade Policy- Difficult Times— The 1970s and Early 1980s.* Te Ara - The Encyclopedia of New Zealand. [Online] Available at: https://teara.govt.nz/en/overseas-trade-policy/page-5 [Accessed 3 September 2021].

O'Connor, P., 2001. Keeping New Zealand White, 1908–1920. In: J. Binney, ed., *The Shaping of History*. Wellington: Bridget Williams Books, pp. 285–307.

Orange, C., 2015. *The Treaty of Waitangi*. 3rd ed. Wellington: Bridget Williams Books, p. 272.

Pretes, M. and Gibson, K., 2008. Openings in the Body of 'Capitalism': Capital Flows and Diverse Economic Possibilities in Kiribati. *Asia Pacific Viewpoint*, 49(3), pp. 381–391.

Sin, I. and Ormsby, J., 2018. *The Settlement Experience of Pacific migrants in New Zealand: Insights from LISNZ and the IDI.* Motu Working Paper. Wellington: Motu Economic and Public Policy Research.

Sofield, T.H.B., 2002. Outside the Net: Kiribati and the Knowledge Economy. *Journal of Computer-Mediated Communication*, 7(2) Available at: https://doi.org/10.1111/j.1083-6101.2002.tb00144.x [Accessed 1 September June 2021].

Spoonley, P., 2015. A Political Economy of Labour Migration of New Zealand: Official Newsletter of the New Zealand Demographic Society, *New Zealand Population Review*, 41, pp. 169–190.

Spoonley, P. and Bedford, R., 2012. *Welcome to Our World?: Immigration and the Reshaping of New Zealand*. Auckland: Dunmore Publishing.

Stats NZ, 2019. *New Zealand's Population Reflects Growing Diversity.* [Online] Available at: https://www.stats.govt.nz/news/new-zealands-population-reflects-growing-diversity [Accessed 3 September 2021].

Stats NZ. n.d. *2018 Census Ethnic Group Summaries.* [Online] Available at: https://www.stats.govt.nz/tools/2018-census-ethnic-group-summaries/pacific-peoples [Accessed 3 September 2021].

The Office of Ethnic Affairs, 2013. *Language and Integration in New Zealand*. Te Tari Taiwhenua | Department of Internal Affairs. [Online] Available at: https://www.ethniccommunities.govt.nz/assets/Resources/7d40a0074e/LanguageandIntegrationinNZ.pdf [Accessed 6 September 2021].

Thompson, M., 2015. *Settlement Experiences of Kiribati Migrants Living in New Zealand*. Ph.D. University of Otago.

Thornton, F., 2011. Regional Labour Migration as Adaptation to Climate Change?: Options in the Pacific. In: M. Leighton, X. Shen and K. Warner, ed., *Climate Change and Migration: Rethinking Policies for Adaptation and Disaster Risk Reduction*. Bonn: United Nations University Institute for Environment and Human Security (UNU-EHS), pp. 81–89. [Online] Available at: https://www.auca.kg/uploads/Source%20Pub%20Climate%20C hange%20and%20Migration.pdf [Accessed 3 September 2021].

Tokalau, T., 2021. *Dawn Raids Apology: PM Sorry for 'Hurt and Distress' of Racially Targeted Policy*. Stuff. [Online] Available at: https://www.stuff. co.nz/national/125849685/dawn-raids-apology-pm-sorry-for-hurt-and-dis tress-of-racially-targeted-policy [Accessed 3 September 2021].

UNDP Human Development Report, 2021a. *Human Development Indicators— Kiribati*. United Nations Development Programme. [Online] Available at: http://hdr.undp.org/en/countries/profiles/KIR [Accessed 12 July 2021].

UNDP Human Development Report. 2021b. *Human Development Indica-tors—New Zealand*. United Nations Development Programme. [Online] Available at: http://hdr.undp.org/en/countries/profiles/NZL [Accessed 13 July 2021].

United Nations Committee for Development Policy, 2018. *Committee for Development Policy 20th Plenary Session: Vulnerability Profile of Kiribati (12–16 March, 2018)*. CDP2018/PLEN/6.b. New York, p. 6. [Online] Available at: https://www.un.org/development/desa/dpad/wp-content/upl oads/sites/45/CDP-PL-2018-6b.pdf [Accessed 7 July 2021].

World Commission on Environment and Development, 1987. *Our Common Future*. Oxford, Oxford University Press.

Wyett, K., 2013. Escaping a Rising Tide: Sea Level Rise and Migration in Kiribati. *Asia & the Pacific Policy Studies*, 1(1), pp. 171–185.

Agency and Action: Gender Inclusion in Planning for Climate Change-Induced Human Mobility in Fiji

Betty Barkha

Climate change-induced human mobilities are almost always projected as an issue of the future. However, the climate crisis[1] is no longer a futurist threat and has already displaced millions across the globe. In 2020 alone, 30.7 million people were displaced within borders by weather and climate-related disasters such as tropical cyclones, hurricanes, typhoons, floods, wildfires, landslides, extreme temperatures and droughts (IDMC 2021: 11). As in most crises, the impacts and experiences vary by individual characteristics and intersecting circumstances such as age, ethnicity, socio-economic class, geographical location, disability and gender. Sudden

---

[1] The use of the term climate crisis is subject of contestations and debate; however, I make frequent references to climate crisis, which refers to the cumulative catastrophic impacts of climate change, to highlight the urgency with which responses are required for climate mitigation and adaptation. For further discussions see: Buxton and Hayes (2015), Hirsch (2015), Thornton et al. (2020), Podesta (2019), McDonald (2012).

B. Barkha (✉)
Centre for Gender, Peace and Security, Monash University, Melbourne, VIC, Australia
e-mail: betty.barkha1@monash.edu

N. J. P. Alsford (ed.), *Pacific Voices and Climate Change*,
https://doi.org/10.1007/978-3-030-98460-1_5

89

and slow-onset impacts leading to environmental degradation and increasingly intensifying climate stressors contribute to climate adaptation strategies linked to human movement. Climate change-induced movements of people are broadly referred to as human mobility and have been classified as three distinct forms of movement. The three forms of movement: forced or involuntary displacement, planned relocation[2], which is predominantly state-led; and climate migration, understood primarily as a voluntary movement (The Nansen Initiative 2015; UNFCCC 2010: 5). These disruptions caused by climate change-induced displacements and planned relocation affect marginalised populations disproportionately, irrespective of whether they occur in developed or developing countries.

In the last decade, a growing number of countries have adopted climate change policies and made significant strides in advancing gender inequality. Expediently there is little overlap between the two agendas, and I argue in this chapter that it is critical gender and climate change intersect in a way that is beneficial to the global political economy. I use Fiji as a case study to illustrate, through critical discourse analysis, how gender inclusion is prioritised in the national context of responding to the climate crisis-induced planned relocation and displacement. This paper undertakes critical discourse analysis, through a feminist lens, to closely examine at the extent to which gender has been integrated in existing frameworks on climate change-induced human mobility such as planned relocation, displacement and climate migration. Undertaking a quasi-audit of gender inclusivity in mapping Fiji's response to climate change-induced human mobility mechanisms that examines relevant policies, processes, frameworks and instruments related to three key aspects. The findings and analysis presented in this paper are part of a larger qualitative research undertaken for a doctoral thesis, which is currently underway.

Gender inclusivity, in the context of climate change, refers to gender-diverse inclusion of individuals or community groups that are impacted by the relevant policy or project (True 2016; Vitukawalu et al. 2015). Climate change is gendered, inherently making the processes and responses of climate change-induced mobility gendered (Dankelman

---

[2] Also referred to as managed retreat, resettlement or planned retreat and can be used interchangeably (Hino et al. 2017; Luetz and Havea 2018; Dannenberg et al. 2019). For this paper, I use planned relocation because of its relevance in the case study of Fiji.

2012; Camey et al. 2020). The impacts of climate change affect individuals differently based on their social, political and economic standings. In the case of responding to climate change-induced planned relocation and displacement in Fiji, this varies from community level to national, regional and global spaces. In assessing gender inclusivity, I look for any commitments that have been prioritised to ensure gender representation and participation across relevant spheres of the commitment. To assess climate change and human mobility mechanisms at the Fijian national level I operationalise gender inclusivity to reflect commitments to enhance gender diverse engagement. For example, gender balance commitments, quotas or anything that mandates gender representative engagement across global, regional and national efforts. I also argue that in order to avoid another human mobility crisis it is critical to take a proactive approach in planning for climate change-induced mobility whether voluntary or involuntary.

Impacts of climate change that are driving people away from homes have significant trauma attached to it, particularly for indigenous populations who have a sacred connection to the land, flora and ocean. In the case of Fiji, the *iTaukei* (indigenous) identity is strongly linked to the *vanua* ancestral identity (patrilineal), well-being and sense of belonging. Sensitivity to cultural context refers to exploration and understanding of the multiplicity of complexities that influence climate-induced mobilities such as planned relocation and displacement. A tangible example of this in Fiji is land ownership, whereby approximately 89% of the land is *iTaukei* land and has strong cultural connections linked with traditional land-owning units such as the *Yavusa* or the *Mataqali*[3]. The *Yavusa* and *Mataqali* traditional groupings fall under the confederacies which are known as the *Vanua*, and as previously highlighted this plays a critical role in indigenous relationships with the land and the ocean (Nabobo-Baba 2008; Lewis et al. 2020). For non-indigenous Fijian communities in Fiji, such as Indo-Fijians, Rabians, Rotumans and other ethnic backgrounds the challenges are unique to the *iTaukei* community but do not diminish their sense of loss and grief in the face of the climate crisis. Historical tensions between the two major ethnic (*iTaukei* and Indo-Fijians) groups have been associated with increasing inequalities in the political

---

[3] The *Yavusa* refers to a group of correlated *Mataqali* groups of traditional landowning units that have specific roles under the *Yavusa*. For further clarity see i-Taukei Land Trust Board (TLTB, n.d., 4–5).

and economic realm, which resulted in smaller ethnic groups (such as the Rabians, Rotumans, Chinese and other Pacific islanders) being excluded from major social, economic and political spaces. Access to and control over resources such as land are also strongly linked to ones' identity and continue to reinforce social and gendered power hierarchies.

This paper recognises the multicausality of climate change-induced human mobility, gender inequality and places emphasis on responding to the climate crisis urgently but simultaneously ensuring that adaptation responses are gender-inclusive and just. In doing so I methodologically use a critical discourse analysis to illustrate the extent of gender inclusion within written and spoken language on climate change in the case of Fiji and how it links to pre-agreed commitments on climate change and gender equality globally. In the following section I pay specific attention to how marginalised communities are affected by precarious impacts of climate change; how the Fijian government responds to these challenges; who makes the decisions and who benefits the most from these processes. By drawing attention to challenges around meaningful gender inclusion in responding to impacts of climate change in the case of Fiji, I advocate for accelerated implementation efforts in gender mainstreaming across all state-led adaptation and mitigation efforts. I argue that responses to climate change impacts can be an opportunity to transform livelihoods, safeguard human security and reduce socio-economic inequalities, such as gender inequality.

Gender and Climate Change-Induced Human Mobilities

Complexities around climate change-induced displacement, planned relocation, resettlement or environmental migration stem from gaps in existing legal and international frameworks that do not recognise the multicausality of human mobility in the face of the climate crisis. Millions across the world migrate regularly for numerous reasons motivated by strong political, social, economic and security pressures. As in any crisis, the impacts of the climate change crisis are gendered, affecting men, women and gender non-conforming[4] communities differently. To distinguish clearly and solely which of these movements is motivated by impacts of climate change, natural disasters or other environmental factors are extremely multi-causal and complex. In a groundbreaking (non-binding) ruling by the United Nations Human Rights Committee it is now unlawful for governments to return people to countries where their lives are threatened by impacts of the climate crisis (United Nations 2020). This was brought to global attention after a Kiribati national challenged his deportation from New Zealand on the basis that impacts of climate change such as sea-level rise, lack of safe drinking water and rising violent conflicts endangered his life (Wasuka 2020). The committee in principle recognised that environmental degradation and climate change constitute serious livelihood threats and will continue to increase in years to come, but ruled against the petitioner (United Nations 2020). The gendered implications of such a ruling are yet unknown; however previous case studies[5] on labour migration and displacement due to violent conflict show that the burden continues to fall on women, girls and gender non-conforming populations.

Pacific SIDS leaders and civil society organisations have been clear and consistent with their messaging on climate change through national, regional and global processes—"*Urgent Climate Action Now*" (Pacific Islands Forum 2019). Additionally, the former President of Kiribati also brought to global attention the concept of *migration with dignity*; however, the current leadership of Kiribati has reprioritised to focus on internal adaptation efforts (Weber 2015; Voigt-Graf and Kagan 2017). The concept of migration with dignity would entail migration with the assurance of protection of human rights to all, including an adequate standard of living and the right to housing, if this concept were to be further explored within regional and global contexts. A 2018 World Bank report highlights that imminent threat to human security resulting from climate impacts (such as droughts, floods, rising sea levels and decreasing crop yields) could push more than 143 million people from their homes

---

[4] This paper uses gender non-conforming to refer to individuals who do not identify with binary gender identities of feminine or masculine. This term is inclusive of lesbian, gay, bisexual and transgender, unless otherwise stated and was a preferred term by participants of the research.

[5] See also: Kronsell (2018), McLeman and Gemenne (2018), Ferris (2019), Tanyag and True (2019); Bertana (2020).

(Rigaud et al. 2018). Currently there is no global framework that comprehensively addresses the challenges of climate change-induced mobility and how these will impact existing social and economic inequalities. Amidst global pressures to address human mobility on a global scale the United Nations introduced the Global Compact for Safe, Orderly and Regular Migration (hereafter referred to as Global Compact), which was the first-ever negotiated global framework on migration that recognizes that migration in the context of disasters and climate change (United Nations 2018).

The Global Compact is grounded in cross-setting guiding principles which emphasise that it is people- centred, based on international human-rights principles and is gender-responsive. The Global Compact also calls for gender mainstreaming and recognition of gender diverse agency, protection of human rights and empowerment of women in related processes. Specifically stressing that "The Global Compact ensures that the human rights of women, men, girls and boys are respected, that their specific needs are properly understood and addressed and that they are empowered as agents of change" (United Nations 2018: 5–6). Given that the Global Compact is a new agreement, its potential remains unexplored at this stage, but it remains a crucial tool in paving pathways for human mobility in the face of the climate crisis.

In early studies of climate-induced mobilities in the Pacific Prof. Jane McAdam emphasised that *"There is so much more to relocation than simply securing territory"* (McAdam 2010). In the Pacific, adverse impacts of climate change (both sudden and slow-onset disasters) threaten livelihoods and are driving factors of human mobility as it continues to destroy existing ecosystems and make entire islands uninhabitable. Degradation of the environment leads to scarcity in essential resources and hence pushes families and communities to make difficult decisions. The case of Carteret Islands in Papua New Guinea was the first extensively broadcasted incident of climate change-induced planned relocation in the Pacific (Edwards 2013; Pascoe 2015). The move has been particularly problematic and traumatic for the Carteret Islanders as they were a matrilineal society in which land ownership was primarily passed through to women (Pascoe 2015: 79). This led to loss of cultural connection with ancestral land and required adjustment in a new location which was already burdened with challenges of violent conflict. Similar accounts of loss, damage and devastation were experienced by residents of Alaska natives from Kivalina and Shishmaref (Marino 2012; Bronen and Chapin

2013), L'Aquila residents in Italy (Guadagno 2016), the Mekong Delta (Chun 2015), Taro Islanders in the Solomon Islands (Haines 2016; Benintende 2019). Planned relocations are major undertakings and can be a transformative tool for not only protecting people and improving their quality of life but also opportunities to safeguard human security and reduce socio-economic inequalities, as has previously been suggested by many scholars[6].

After the early mention of climate change-induced mobilities was made through the United Nations Framework Convention on Climate Change (UNFCCC) Cancun Adaptation Framework (2010 para. 14(f)); recent mechanisms such as the Global Compact for Migration (United Nations 2018), the Paris Agreement (United Nations 2015), Agenda for the Protection of Cross-Border Displaced Persons in the Context of Disasters and Climate Change (The Nansen Initiative 2015) and Sendai Framework for Disaster Risk Reduction (United Nations Office for Disaster Risk Reduction 2015) have increasingly devoted attention to human mobilities in connection to adverse impacts of climate change. While the impacts of climate change globally will force human mobility, it is critical that this not be viewed as a failure of communities to adapt but rather a form of adaptation. Till date, climate change-induced mobility such as planned relocation and migration have not been acknowledged in positive light (Ash and Campbell 2016; Hermann and Kempf 2017). Decisions around climate-induced mobilities remain complex, and experiences are different for women, men and gender non-conforming individuals.

At the 21st Conference of Parties (COP) to the UNFCCC a task-force on displacement was created to specifically look at issues arising from climate change mitigation and adaptation induced human mobility (Heslin et al. 2019). In 2015, the Nansen Initiative (2015: 16–17) endorsed a protection agenda for people displaced by climate change and noted that human movements as a result of climate change impacts might be more of coercive than a choice. In all cases of climate-induced movements of people there is significant challenges posed to human security, self-determination, identity, culture, loss of home and adapting to the ways of host communities (McAdam 2014; Kothari and Arnall 2019). However, the degree to which people are affected by climate change-induced mobilities is heavily linked to their gender, age, socio-economic

---

[6] See: Hugo (2011), McAdam (2013), Ahmed (2018), Luetz and Havea (2018), Barnett (2020), Ferris and Weerasinghe (2020).

status, power over and access to resources. While all mechanisms mention the gendered disparities to some extent, there was limited effort made to address it holistically until recently under Gender Action Plan commitments to UNFCCC.

Gender issues took some time to enter in climate change discourse, initiatives and processes but has evolved greatly since COP 6 held in The Hague in 2000, after which gender mainstreaming became a key point of debates (Dankelman 2002: 25). The Hague conference was also monumental in mandating the Adaptation Fund that would provide specific guidance to SIDs through the Global Environment Facility. This was followed by a resolution proposed by the Samoan delegation at COP 7 in Marrakech, calling for enhanced participation of women at climate change negotiations which was later adopted as an action decision for state parties (UNFCCC 2002). The decision specifically called for "Improving the participation of women in the representation of Parties in bodies established under the UNFCCC or the Kyoto Protocol" (UNFCCC 2002: 26). While gender balance entirely based on numbers does not guarantee a gender-inclusive outcome and presence does not necessarily imply power, it is a step in the right direction in bridging the gender gap in climate governance mechanisms.

Another notable change has been the mandating of gender assessments and gender action plans across all projects that require resourcing from the Green Climate Fund (GCF). The GCF was established by UNFCCC state parties in 2010 to act as the Convention's financial mechanism (Green Climate Fund 2021; UNFCCC 2010). To support national-level integration of gender equality in climate change interventions and climate finance, the GCF has initiated numerous decentralised capacity development initiatives, accompanied by a manual with clear targets and indicators on gender mainstreaming (Green Climate Fund 2017). Despite concerns around inaccessibility to GCF for most Pacific nations[7], this provides states and accredited partner entities to holistically mainstream gender in GCF-supported projects and provides a toolkit to guide implementation efforts. The GCF toolkit on gender mainstreaming

---

[7] Many Pacific SIDs expressed disappointment over the complex procedures required to access GCF funding. This required states to rely on third party partners such as the Australian government and Secretariat of the Pacific Community to facilitate funding from GCF (Maclellan and Meads 2016).Query Currently 14 Pacific SIDs can access funding from the GCF (Green Climate Fund 2021).

emphasises the disproportionate impacts of climate change on women and girls, recognises the role of social reproduction in responding to the climate crisis and calls for gender mainstreaming across the project cycles (Green Climate Fund 2017). The GCF Gender mainstreaming manual was developed in collaboration with UN Women and provides templates on how to integrate gender across project cycles, however recognises that context matters, and it needs to be customised accordingly. Taking a gender-sensitive approach, which recognises the differentiated impacts and agency of men, women and gender non-conforming individuals is crucial in co-addressing gender inequality while acting on climate change (Green Climate Fund 2017). While there is no singular approach to mainstreaming gender, there are common indicators and targets that states can prioritise within national adaptation and mitigation efforts.

Responding to gender inequality, safeguarding of human security and violation of human rights have become a significant factor in contemporary geo-political spaces. The adoption of key instruments like the Convention of the Elimination of All Forms of Discrimination Against Women (CEDAW), Beijing Platform for Action (BPfA), the Women, Peace and Security Agenda (WPS) and the Sustainable Development Goals (SDGs), have incentivised states and institutions to prioritise gender mainstreaming efforts. Similarly, in the climate change arena, the implementation of Gender Actions Plans (GAP) has become a key priority for states that have proven to be crucial tools in advancing action on gender equality (Asian Development Bank 2020; Seymour 2020: 25).

It was only at COP 25 in 2019 that state parties agreed to adopt and report on gender action plans within the context of their climate commitments (UNFCCC 2019a). At COP 25, state parties agreed to an enhanced Gender Action Plan (GAP) that prioritises five key areas to advance gender-responsive climate action and gender mainstreaming in the implementation efforts. The five key areas are: (a) capacity building, knowledge management and communication (b) gender balance, participation and women's leadership (c) coherence (d) gender-responsive implementation and means of implementation and (e) monitoring and reporting (UNFCCC 2019a, b: 6–8). Within UNFCCC and its linked mechanisms, the Warsaw Mechanism and Cancun Agreements also provide pathways for gender mainstreaming, particularly linked to climate change-induced human mobility. Action area 1 of the rolling five-year Warsaw Mechanism workplan was to: "enhance the understanding of how loss and damage associated with the adverse effects of climate change affect particularly vulnerable developing countries, segments of the population that are already vulnerable owing to *geography, socioeconomic status, livelihoods, gender, age, Indigenous or minority status or disability, and the ecosystems* that they depend on, and of how the implementation of approaches to address loss and damage can benefit them" (UNFCCC 2013, 2019b). Whereas Action area 6 looks to "enhance the understanding of and expertise on how the impacts of *climate change are affecting patterns of migration, displacement and human mobility;* and the application of such understanding and expertise" (UNFCCC 2019a). How this manifests in local and national level climate responses is an area of growing investigation, particularly in research on Loss and Damage.

The Fijian government is heavily invested with the Warsaw Mechanisms and has supported their workplan by hosting the Fiji Clearing House for Risk transfer[8] and the Suva Expert Dialogue on Loss and Damages associated with Climate Change. The action items and progress of the Warsaw Mechanism also show that it can be a crucial tool for mainstreaming gender within the global climate change-induced human mobility. The Warsaw Mechanism is now uniquely positioned within UNFCCC bodies to engage and strengthen financing mechanisms to developing countries and to provide thematic experts to support technical guidelines and implementation plans, making it a key player in mainstreaming gender within UNFCCC mechanisms (UNFCCC 2019b, 2021).

Fiji has, and continues to be, a key player in the global political economy for the last two decades, it continues to show leadership at a global level through commitments on climate change and gender equality. Some key leadership roles include Fiji's chairing of the Platform on Disaster Displacement for the term 2021–2022 (Platform on Disaster Displacement 2021), President of UN Human Rights Council for the 2019–2021 term, President of the UN Climate Conference, COP 23 and President for the UN General Assembly in 2016. Fiji's leadership within global platforms was a significant step in globally advocating for climate change as the greatest crisis of our time and a major threat to international peace and security (United Nations 2019). However, whether actions align with words is questioned, for example, the Fiji's current Prime Minister has been widely acknowledged as a champion for climate change

---

[8] The Fiji Clearing House for Risk Transfer was established to serve as a repository for information on climate insurance and risk transfer linked to Loss and Damage. See UNFCCC (2019a), Ourbak and Magnan (2018), Byrnes and Surminski (2019).

(UNEP 2020) and much of global advocacy and leadership has come through since his takeover of the Fijian parliament in 2006. However, often his words are contradicted by his actions nationally when it comes to addressing gender equality and valuing women[9]. Moreover, the PM also received backlash when he publicly stated that those seeking same-sex marriage in Fiji should move to Iceland, because there is no place for homosexuality in [the Christian state of] Fiji (DIVA for Equality 2019: 50; Sheldrick 2016).

Aligned with feminist scholars in International Relations, I also argue that existing social, economic and political structures are engrained in practices of neoliberalism and patriarchy. As Lee-Koo has previously emphasised, often when responding to crises (whether climate or conflict) states lose an opportunity to address gender inequality by dismissing it as not urgent (Lee-Koo 2018). Gender diverse inclusion is critical in responding to the climate crisis in a sustainable, equitable and efficient way. For example, an increase in women's participation at the political level has resulted in greater responsiveness to citizen's needs, often increasing cooperation across party and ethnic lines and delivering more sustainable and peaceful outcomes (Asia Pacific Forum on Women 2015; WEDO 2020). The following section dives deeper into Fiji's development and execution of the Planned Relocation and Displacement Guidelines and examines how gender issues have been integrated within responses.

Fiji's Experience of Climate-Induced Planned Relocation and Displacement

In the Fiji Islands, the government has fully and partially relocated five villages and identified over eighty that may need future relocations[10] (Ministry of Foreign Affairs and International Cooperation 2014: 136). The Fijian government's understanding of planned relocation is that it is

[9] For example, PM Bainimarama has been noted telling Opposition member, Lynda Tabuya, to cover up during a parliamentary session (Vakasukawaqa 2019) and pointed at former Opposition Leader, Ro Teimumu Kepa's face while telling her that she needed a husband if she wanted match his travel allowance (Rika 2018). He also undermined professional merits targeted at another Opposition member's wife, which received extensive backlash from civil society organisations (Kate 2021).

[10] Villages that have been supported by the Fijian government through full or partial planned relocation: Vunidogoloa, Denimanu, Vunisavisavi, Narikoso, and Tukuraki.

an option of last resort and is a state-led initiative directed at physically moving communities to safer ground (Ministry of Economy 2018: 6–7). Displacements on the other hand refer to human mobility as a result of either violent conflict, violation of human rights or sudden natural or human-made disasters that forces people out of homes and is not state-led (Fiji Ministry of Economy 2019: 5). This section examines the Fijian Governments National Climate Change Policy (Ministry of Economy 2019), Displacement Guidelines (2019), Planned Relocation Guidelines (2018), 5-Year and 20-Year National Development Plan (2017) and Fiji's National Gender Policy (Ministry for Social Welfare 2014).

Despite contributing negligibly to global carbon emissions, Fiji has conditionally committed to curbing emissions from the energy sector by 30% and generating 100% of its electricity from renewables by 2030. Achieving these ambitious development goals will require unprecedented levels of finance to be mobilised and channelled into sustainable development projects rapidly. Meeting national emissions reduction targets in the energy sector alone in Fiji will require an estimated USD $2.97 billion between 2017 and 2030 (Government of Fiji 2017, 2019). Since 2011, Fiji has fully and partially relocated *iTaukei* villages devasted by impacts of climate change and was left with no choice but to move. Climate change-induced mobility efforts are expensive and economically draining. The costs of climate change adaptation and induced mobilities are much higher than just economic and financial losses. The World Bank estimates that adaptation initiatives will cost Pacific SIDS significantly, for example coastal protection efforts alone might cost the Marshall Islands USD $58 million (13% of GDP) and Fiji USD $329 million (3% of GDP) per year (The World Bank 2017: 4). These projections were based on the best-case scenario and did not account for other adaption efforts or disaster recovery and response initiatives. In future years, the Pacific region is projected to receive substantial flows of climate finance, however how these are targeted to holistically incorporate gender mainstreaming within climate projects remains unknown.

It is important to emphasise that Fiji remains governed by an authoritarian administration which came into power initially through a coup which was later legitimised through elections (Ratuva and Lawson 2016). This becomes a critical area of concern, particularly because militarised and authoritarian regimes mobilise a gendered form of leadership that is hardly concerned with holistically transforming gendered disparities. In

Fiji's case this may be disputed given the advancements made nationally towards progressing gender equality commitments. Moreover, this mapping of gender inclusivity is vital to understand the frameworks and instruments that already exist, their gaps regarding gender, and whether they are sufficient to address the gendered impacts of state-led responses to climate change-induced planned relocation or displacement. Existing mechanisms that need to integrate and address gender issues while responding to climate, while progressive, are insufficient to address underlying power imbalances which inherently reproduce gender inequality. Feminist scholars and human geographers studying climate change have argued that structures and individuals within key political, social and economic institutions continue to harness gender dynamics to retain or further existing power, status and authority within institutions (Adger 2000; Aggestam and True 2021). This is particularly relevant in the case of state-led and donor-supported responses to climate change, as institutions are structures of power and inherently gendered as is evident in the case of Fiji.

The first Fijian planned relocation in Vunidogoloa came with 30 new homes, a copra drier, pineapple farm, cattle and fishponds. Each family received a new home, equipped with solar lights, and a flush toilet. School children were no longer missing school because of flooding. This infrastructure was intended to support the transition of villagers further uphill by providing them options for livelihood support in a new location. However, some of the new homes came without kitchens and regular electricity leaving the villagers to rely on firewood and kerosene energy for cooking, adding to the workload of women within the village (Bertana 2020; Piggott-McKellar et al. 2019a; Singh et al. 2020). The absence of women throughout the planning processes led to some homes being built without kitchens with poor quality leftover materials and low lighting that affected women's ability and efficiency in doing everyday tasks. Additionally, at the household level, women were primary caretakers and caregivers, which involves the undertaking of reproductive or care activities that affect the well-being of both current and future generations such as food preparation, domestic work, subsistence production, childcare, care for the sick and elderly, collection of fuel and water, etc., which are mainly performed using unpaid labour in the household.

Additionally, evidence that links increasing intimate partner violence and gender-based violence during and post-natural disasters such as forced displacement during cyclones or floods, the Fiji government continues to limit funding for crisis response or domestic violence support (UN Women 2014; Republic of Fiji 2021: 156). This significant under-resourcing of violence response support mechanisms continues to hinder full protection of women and gender non-conforming individuals in the face of the climate crisis. Since 2011, Fiji has fully and partially relocated six communities devasted by impacts of climate change and were left with no choice but to move and has identified up to 43 more villages that need to be relocated in the near future (Ministry of Economy 2020).

The two key documents guiding the Fijian government's responses on climate change-induced mobility are the Planned Relocation Guidelines (2018) and the Displacement Guidelines (2019). The Planned Reloca-tion and Displacement Guidelines remain living documents without an active operating manual (in this thesis referred to as *Standard Operating Procedures* or SOPs) that has been in development stages since 2018 (Corendea 2016). The Planned Relocation Guidelines are intended to provide guidance for the Fijian Government and relevant stakeholders, when considering planned relocation within Fiji as an option of last resort for climate adaptation efforts. The guidelines facilitate state-led responses in the Fijian context and are grounded in a human-rights-based and livelihood-centred approach across decision, planning and implemen-tation phases. Lessons from Fiji's experiences with planned relocation processes pre-guidelines where community movements have been asso-ciated with numerous social, cultural, economic and environmental issues are intended to be avoided using the Planned Relocation Guidelines and accompanying operating procedures (Ministry of Economy 2018: 8). Tensions over displacement from traditionally owned land to new sites, inadequate resources, improper communication and incomplete infras-tructure burdened by lack of psycho-social support will now inform the development of Standard Operating Procedures.

The Displacement Guidelines begin by clarifying that forced displace-ment of people because of climate change impacts and natural disasters is not a state-led process, however, requires state response. They are risk aversive in essence placing emphasis on preventative measures that could avert or minimise impacts of forced displacement on human secu-rity. Building on the principles of being human-centred, livelihood-based and human-rights-based, the displacement guidelines also emphasise approaches to capacity building and equity (Ministry of Economy 2019, 10–11). Given that the displacement process is involuntary in nature, the responses emphasise stages pre-, during and post-displacement that need to be targeted at safeguarding human security through a gender-responsive consultative and participatory process. Both the Planned Relocation and Displacement guidelines place emphasis on the State's responsibility and duty to protect citizens and ensure their rights and security are guaranteed. However, without clear SOPs both sets of guidelines remain largely speculative.

The Planned Relocation Guidelines, in its commitment to a livelihood-based approach, recognises that the planned relocation process needs to be sensitive to the specific needs of communities and households that may be on the move (Ministry of Economy 2018: 8). This amplifies the need for an intersectional approach that is mindful of rights belonging to the affected people, regardless of age, ethnicity, faith, elderly, disabilities and gender are meaningfully engaged and able to participate in the decision-making, planning and implementation of planned relocation efforts. In undertaking a state-led planned relocation effort, the burden falls upon implementation teams to ensure that a reactive approach is also considered with a pre-emptive approach to avoid further escalation of crises during the process of planned relocations. Both the Planned Relocation and Displacement guidelines reiterate identical commitments to being inclusive and gender-responsive across various phases of state-led responses to climate change-induced human mobility in Fiji. The commitments are vital to the consultative and participatory processes underlined within the guidelines. Within the normative frameworks of the Planned Relocation and Displacement Guidelines, overarching principles are suggested for pre-, during and post planned relocation or forced displacement. Within the Displacement Guidelines a similar obligation is highlighted with the human-rights based approach which require sensitivity across intersections of age, gender that is aligned with the rule of law (Ministry of Economy 2019: 10).

However, even in the absence of comprehensive SOPs, there are alternative pre-agreed frameworks and national policies that can be utilised to address gendered disparities. For example, Fiji's National Gender Policy (Ministry for Social Welfare 2014) developed gender-sensitive indicators aligned with global and regional commitments for national adoption since 2014. The indicators range from promoting international gender-sensitive standards across institutions to recognising gendered roles in social reproduction and encouraging institutional change to accommodate evolving gender roles in public and private spheres. The National Gender Policy recognised the need for flexible working hours, increased access to quality childcare facilities, flexible work arrangements, cost sharing and subsidised access to services for lower income families (Ministry for Social Welfare 2014: 16). The National Gender Policy outlines a clear framework for ensuring gender sensitivity in all state-led activities while recognising the gendered disparities based on physiological, biological and social differences in the Fijian context. Research shows that meaningful gender inclusion in climate governance provides necessary insights that embody social equity, reflect and serve the needs of communities. For example, a 2019 research shows that greater women's representation in national parliaments is likely causally connected with stronger climate policies (Mavisakalyan and Tarverdi 2019; Tanyag and True 2019).

In addition to the Planned Relocation and Displacement Guidelines, the National Climate Change Policy (NCCP) of Fiji (2018–2030) also weaves itself within three policy pillars of being human-rights-based, evidence-based and gender-responsive. Within the policy pillar of gender responsiveness, the document recognises women as agents of change, acknowledges the need for social adaptation to address inequalities, distinguishes the gendered impacts of climate change, takes a risk-aversive approach and is informed by gender-sensitive indicators.[11] In the case of Fiji, monitoring gender mainstreaming across project cycles depends directly on the resources dedicated towards planning, implementation and evaluation efforts. A critical factor in the global and local responses to climate change has been climate financing. Pacific SIDS are some of the most vulnerable countries on the frontlines of the climate crisis, which pulls up a hefty annual bill. Climate resourcing is specifically needed to address the gender inequalities exacerbated increasing impacts of climate change in Fiji.

Gendered gaps in climate governance need to be addressed through equitable and meaningful participation across gender diversities. This includes men and boys, women and girls and gender non-conforming individuals. Women, gender non-conforming individuals, children and other persons such as persons with disabilities, elderly, indigenous peoples and migrants should not be seen only as victims or in terms of vulnerability. They should be recognised as agents of change, educators and essential partners in the local, national and international efforts to tackle

---

[11] These include but are not limited to sex disaggregated data on households and individuals benefitting from adaptation projects.

climate change. Enabling gender and socially diverse voices in decision-making can accelerate lasting transformational solutions for sustainable development, shaping equitable policies and resource allocation. Moreover, reports show that people with disabilities and diverse SOGIESC[12] are often neglected from response efforts and their priorities, while now increasingly being prioritised through inter-agency taskforces, remain overlooked (DIVA for Equality 2019; Samuwai and Fihaki 2020). They rarely receive any information through community members, which further limits their agency and decision-making powers. In the following section I elaborate on aspects of social and cultural barriers that continue reinforce gender inequality within the Fijian context, which need to be addressed holistically in response to climate change-induced planned relocations and displacement.

Socio-Cultural Boundaries

In this section I elaborate on socio-cultural aspects which impact gendered experiences of climate change-induced human mobility in the Fijian context. This includes exploring the connections between national level policies and how they are perceived within local and household settings. I emphasise the importance of social reproduction in sustaining life particularly during a crisis and how national-level plans need to account for gender-inclusive processes meaningfully in order to address existing inequalities and then transform them.

As climate change challenges human security through sudden and slow-onset environment degradation, gender roles evolve to adapt with increasing loss, damage and grief associated with impacts. As a middle-income country, Fijis economy is heavily reliant on tourism, fisheries and agriculture. However, due to increasing frequency and intensity of cyclones, it suffers major losses and damages to the primary sectors (World Bank 2017). TheWorld Bank (2017: 22) estimates an average of 25,700 people are pushed into poverty by impacts of natural disasters and climate change. The gender gap in income and poverty rates globally and in the region inherently places women, youth and gender non-conforming individuals at higher risk of economic insecurities (Kabeer 2015; True 2012). Additionally, women across Asia and the Pacific are burdened with social

---

[12] SOGIESC—sexual orientation, gender identity and expression, and sex characteristics.

reproductive roles by doing at least four times more unpaid care work than their male counterparts (International Labour Organization 2018). In a study across areas with higher prevalence of poverty in Fiji, data showed that 82 per cent of women spent majority of their time cooking, cleaning, washing clothes, looking after children or other family members and fetching water or cooking fuel (Fisk and Crawford 2017).

Social reproduction and care are vital elements influencing gendered norms and practices that are accepted as normal in Fijian households. The time spent on paid work and unpaid work in households and communities is a crucial element in understanding value, through an expanded lens of money, labour and time. Women are primary care providers in Fiji, disruptions such as forced displacements, planned relocations and ad-hoc community consultations have a significant impact on mothers as they face increased pressure to meet domestic responsibilities, especially in terms of food security and accessing necessary childcare. Increasing frequency and intensity of sudden events such as flooding, droughts, cyclones have significant social, economic and psychological impact on Pacific SIDS communities. The negative impacts of both are felt differently by communities and individuals, depending on a variety of factors, however these challenges have created new entry points for transformation of gender roles and power relations in local settings. For instance, families previously relying on primary industries such as agriculture or fisheries or tourism[13] for income have chosen other forms of employment in urban centres causing an increase in rural to urban drift within and across borders (Clement et al. 2021).

Integrated action to respond to the gender dimensions of climate-induced planned relocation and displacement is important to minimising threats to peace and security, socio-economic inequality and tensions over natural resources access, use and control. For example, the process of partially relocating homes in Narikoso village took over 8 years to negotiate between *mataqali's* and government officials (Barnett and McMichael 2018; Simpson 2020). It was hard to go past the fact that villagers were told by the Prime Minister that it would be best to partially relocate the village to higher ground, which required negotiation with another *mataqali* and did not go satisfactorily (Jolliffe 2016). Another factor that needs further exploration is the fact that the village chief was a

---

[13] This is also largely due to the decline in international travel amidst the global COVID-19 pandemic.

woman and whether this played a role in the arising complexities remains unknown for now.

Access and control over primary resources such as ownership of land are often through patrilineal descent, however changes to the Native Land Act now guarantee women their birthright as landowners and members of the *mataqali* (Jalal 2008; Kopf et al. 2020). However, as a result of culturally constructed gender norms and power gaps, unequal access, control and decisions around the land remain in the hands of the male members of the *mataqali*. The loss of land due to sea-level rise or land-slides is environmental depletion and associated with ties of kinship to traditional landholding units. Inherently these systems of land ownership are gendered and perpetuated by intersecting socio-cultural practices such as religion, kin, age and socio-economic class (Meo-Sewabu 2016: 99).

Evidence also suggests that planned relocations, if planned and financed appropriately, have the potential to deliver positive outcomes that are aligned with global development commitments in reducing inequal-ities and improving livelihoods (Drinkall et al. 2019). For example, the relocation of the coastal village of Vunidogoloa in Fiji further inland reduced their exposure to coastal erosion, has improved accessi-bility, infrastructure, sanitation and livelihood options for the community (Piggott-McKellar et al. 2019b: 9). However, the women felt their unpaid domestic labour roles had increased during and after the project concluded (Bertana 2020, 11). Bertana's study also showed how local and traditional structures continued to limit women's input from decision-making spaces. "*You are not told that you cannot attend, but if you are not specifically asked to attend then you do not go*" (Bertana 2020, 12). Planned relocations efforts can have positive outcomes if they are informed by gender analysis and undertake a human right centred participatory approach.

Culturally sensitive ways need to be prioritised in order to engage effectively with affected communities. This includes open and transparent *Talanoa* sessions which allow community members to freely express their sentiments of climate change impacts, which includes emotions of loss, grief and trauma. Processes also need to expand to account for diverse ethnic and gender groups and their cultural and traditional practices. For instance, it is well established that religion is a key influencing factor for many in the Pacific, and disruptive processes such as planned relocations need to be in collaborative partnerships with religious and community leaders. Recognising that religious and cultural protocols in Fiji are also based on gendered norms which continue to shape everyday interactions and reinforce gendered power dynamics George (2015, 2019). The Fijian government projects have strategically been working closely with religious leaders, donors and community members to learn from their experiences and strengthen the Standard Operating Procedures which will continue to guide future relocation efforts in Fiji. The close collaboration with cultural and religious leaders in addressing climate change are also an oppurtunity to transform existing social and gender based inequalities.

Gender inclusivity is a singular component within the larger gender mainstreaming ecosystem. Therefore, to ensure mechanisms are adequately addressing gendered disparities there needs to be a thorough assessment to map out contextually relevant issues. For instance, ensuring meaningful gender representative participation is a start, but the processes need to be two-tiered and account for urgent immediate action while ensuring longer term sustainability. A tool that has accelerated national-level gender mainstreaming in Fiji is the Green Climate Fund's (GCF) requirement to include gender assessments and plans for gender mainstreaming within project proposals (Green Climate Fund 2018). Specifically looking at the representation of women and gender non-conforming individuals within decision-making spaces, informed and prior consent for community members, timeliness of community consultations and whether mandatory gender-sensitive and inclusive indicators, as mandated in the National Gender Policy (2014) for all state initiatives, are complied with.

"We need to arm ourselves with the ability to act now. We cannot wait for communities to be drowned out by the encroaching tides. We need a holistic approach; we need adequate resources, and we need it now" (Bainimarama 2019). State-led responses to climate change-induced planned relocation and displacements need to account for economic and non-economic resources to ensure resources are secured prior to community engagements. Unless stringent measures are enforced within the Standard Operating Procedures to comply with contextually relevant gender inclusion measures, it will remain a tick the box exercise. To meaningfully be gender inclusive, active efforts must be made across all stages of the project cycle to involve community representatives in a consultative and participatory approach, with capacity development support if and as needed (Barclay et al. 2019). Overall, commitments under the *Paris Agreement Gender Action Plans, Warsaw International Mechanism for Loss and Damage, and the Cancun Agreements* show encouraging commitments towards climate change-induced mobility responses being gender-sensitive, gender-inclusive and gender-responsive. In the Fijian context, there are positive signs in community consultations, which now no longer occur without women, youth or other groups historically marginalised in traditional settings (Bertana 2020; Farbotko 2020). This assertion is corroborated by discourse analysis of key texts on climate change-induced mobility cited within this research.

Conclusion

In this paper I sought to analyse the extent to which gender is conceptualised and integrated within the complex processes of climate change-induced planned relocation and displacement in Fiji. My aim was to illustrate how gender is conceptualised and examine how Fiji's climate adaptation responses will address gender inequality within local and national settings. Global and regional efforts need to strengthen mandates around gender mainstreaming because of the role of multilateral systems in programming and funding climate adaptation and mitigation interventions. Localisation of global commitments is equally important because unless they are targeted at transforming gender and social inequalities, they will continue to reinforce harmful cycles embedded within patriarchal practices. It goes without saying that by promoting meaningful gender inclusion and harnessing their roles as agents of change, climate response efforts are strengthened.

The challenge, however, is not only to initiate these spaces but also to sustain them over time to allow for intergenerational transitions in responding to climate change. The design, decision-making, and implementation efforts must respect, protect and fulfil the rights of all, including by mandating human rights, safeguarding human security, due diligence and ensuring access to education, awareness raising, environmental information and public participation in decision-making. State-led initiatives particularly have the responsibility to protect and effectively defend the rights of all impacted by the climate crisis.

References

Adger, W. N. (2000) 'Institutional adaptation to environmental risk under the transition in Vietnam', *Annals of the Association of American Geographers*, 90(4), pp. 738–758. https://doi.org/10.1111/0004-5608.00220.

Aggestam, K. and True, J. (2021) 'Political leadership and gendered multilevel games in foreign policy', *International Affairs*, 97(2), pp. 385–404. doi: https://doi.org/10.1093/ia/iiaa222.

Ahmed, B. (2018) 'Who takes responsibility for the climate refugees?', *International Journal of Climate Change Strategies and Management*, 10(1), pp. 5–26. doi: https://doi.org/10.1108/IJCCSM-10-2016-0149.

Ash, J. and Campbell, J. (2016) 'Climate change and migration: the case of the Pacific Islands and Australia', *The Journal of Pacific Studies*, 36(1), pp. 53–71.

Asia Pacific Forum on Women, L. and D. (2015) *Climate change and natural disasters affecting women peace and security*. Available at: http://apwld.org/climate-change-and-natural-disasters-affecting-women-peace-and-security/.

Asian Development Bank (2020) *How to use gender approaches to build climate resilience*. Available at: https://www.adb.org/publications/climate-risk-man agement-adb-projects (Accessed: 20 July 2020).

Bainimarama, F. (2019, September 24) 'We need to arm ourselves with the ability to act now. We can't wait for communities to be drowned out by the encroaching tides'.—PM Frank Bainimarama at the launch of Fiji's climate relocation and displaced peoples trust fund for communities and infrastructure. *Cop23*. Available at: https://cop23.com.fj/climate-relocation-and-displa ced-peoples-trust-fund/.

Barclay, K. et al. (2019) *Pacific handbook for gender equity and social inclusion in coastal fisheries and aquaculture*. Noumea.

Barnett, J. (2020) 'Global environmental change II: Political economies of vulnerability to climate change', *Progress in Human Geography*. doi: https://doi.org/10.1177/0309132519898254.

Barnett, J. and McMichael, C. (2018) 'The effects of climate change on the geography and timing of human mobility', *Population and Environment*, 39(4), pp. 339–356. doi: https://doi.org/10.1007/s11111-018-0295-5.

Benintende, E. (2019) *The relocation of Taro Island—The Architectural League of New York*. Available at: https://archleague.org/article/the-relocation-of-taro-island/ (Accessed: 5 May 2020).

Bertana, A. (2020) 'The role of power in community participation: Relocation as climate change adaptation in Fiji', *Environment and Planning C: Politics and Space*, p. 239965442090939. doi: https://doi.org/10.1177/239965442090 9394.

Bronen, R. and Chapin, F. S. (2013) 'Adaptive governance and institutional strategies for climate-induced community relocations in Alaska', *Proceedings of the National Academy of Sciences of the United States of America*, 110(23), pp. 9320–9325. doi: https://doi.org/10.1073/pnas.1210508110.

Buxton, N. and Hayes, B. (2015) 'Introduction: Security for whom in A time of climate crisis?', in Buxton, N. and Hayes, B. (eds) *The secure and the dispossessed*. London: Pluto Press, pp. 1–20. Available at: https://www.jstor.org/stable/pdf/j.ctt18gzdk7.6.pdf?refreqid=excelsior%3A8578693cb5c6eb707964d60c36f2641e.

Byrnes, R. and Surminski, S. (2019) *Addressing the impacts of climate change through an effective Warsaw International Mechanism on Loss and Damage: Submission to the second review of the Warsaw. International Mechanism on Loss and Damage under the UNFCCC.* London: Grantham Research Institute on Climate Change and the Environment and Centre for Climate Change Economics and Policy, London School of Economics and Political Science.

Camey, I. C. et al. (2020) *Gender-based violence and environment linkages: The violence of inequality, gender-based violence and environment linkages: The violence of inequality.* Gland: IUCN, International Union for Conservation of Nature. doi: https://doi.org/10.2305/iucn.ch.2020.03.en.

Chun, J. M. (2015) *Planned relocations in the Mekong Delta: A successful model for climate change adaptation, a cautionary tale or both?* Washington, D.C. Available at: www.brookings.edu (Accessed: 8 July 2020).

Clement, V. et al. (2021) *Groundswell part 2.* Washington DC: World Bank, Washington, DC. Available at: https://openknowledge.worldbank.org/han dle/10986/36248 (Accessed: 1 October 2021).

Corendea, C. (2016) 'Hybrid legal approaches towards climate change: Concepts, mechanisms and implementation', *Annual Survey of International & Comparative Law*, 21, pp. 29–41. Available at: https://digita lcommons.law.ggu.edu/cgi/viewcontent.cgi?referer=&httpsredir=1&article= 1194&context=annlsurvey (Accessed: 28 September 2020).

Dankelman, I. (2002) Climate change: Learning from gender analysis and women's experiences of organising for sustainable development. *Gender & Development*, 10(2), pp. 21–29. doi: https://doi.org/10.1080/135520702 15899.

Dankelman, I. (2012) 'Climate change, human security and gender', in *Gender and climate change: An introduction.* doi: https://doi.org/10.4324/978184 9775274.

Dannenberg, A. L. et al. (2019) 'Managed retreat as a strategy for climate change adaptation in small communities: Public health implications', *Climatic Change*, 153(1–2). doi: https://doi.org/10.1007/s10584-019-02382-0.

Diverse Voices and Action (DIVA) for Equality (2019) *Unjust, unequal, unstoppable: Fiji Lesbians, bisexual women, transmen and gender non conforming people tipping the scales toward justice.* Laucala Beach Estate. Available at: https://drive.google.com/file/d/1D2YiPOQb_erOxBK2rdRt45Z8mEB 1no0z/view?fbclid=IwAR021L4AbNLCUHs1wbLpwGUd_8Qicpcp08ZwS_ xBwWZd-U-ZglU7Vqnj4uk (Accessed: 2 April 2020).

Drinkall, S. et al. (2019) 'Migration with dignity: A case study on the livelihood transition of micronesians to Portland and Salem, Oregon', *Journal of Disaster*

Research, 14(9), pp. 1267–1276. doi: https://doi.org/10.20965/jdr.2019.p1267.

Edwards, J. B. (2013) 'The logistics of climate-induced resettlement: Lessons from the Carteret Islands, Papua New Guinea', *Refugee Survey Quarterly*, 32(3), pp. 52–78. doi: https://doi.org/10.1093/rsq/hdt011.

Farbotko, C. (2020) Is it too late to prevent systemic danger to the world's poor? *Wiley Interdisciplinary Reviews: Climate Change*. Wiley-Blackwell. https://doi.org/10.1002/wcc.609.

Ferris, E. (2019) 'Protecting displaced women and girls', in Davies, S. E. and True, J. (eds) *The Oxford handbook of women, peace, and security*. Oxford University Press, pp. 500–515. doi: https://doi.org/10.1093/oxfordhb/9780190638276.013.38.

Ferris, E. and Weerasinghe, S. (2020) 'Promoting human security: Planned relocation as a protection tool in a time of climate change', *Journal on Migration and Human Security*, XX(X), pp. 1–16. doi: https://doi.org/10.1177/2331502420909305.

Fisk, K. and Crawford, J. (2017) *Exploring multidimensional poverty in Fiji: Findings from a study using the individual deprivation measure*. Melbourne. Available at: https://iwda.org.au/assets/files/IDM-Fiji-Final-Study-Report-31072017.pdf.

George, N. (2015) 'Starting with a prayer': Women, faith, and security in Fiji. *Oceania*. doi: https://doi.org/10.1002/ocea.5078.

George, N. (2019) Climate change and 'Architectures of Entitlement' beyond gendered virtue and vulnerability in the Pacific Islands? in Kinnvall, C. and Rydstrom, H. (eds) *Climate hazards, disasters, and gender ramifications*. London: Routledge, 1st ed., pp. 101–121.

Government of Fiji (2017) *5-Year & 20-year national development plan: Transforming Fiji*. Available at: https://cop23.com.fj/wp-content/uploads/2018/03/5-Year-and-20-Year-National-Development-Plan.pdf (Accessed: 28 September 2019).

Government of Fiji (2019) 'Fiji focus'. Available at: https://www.fiji.gov.fj/getattachment/0227457e-99d7-42f4-bf2c-0e51a124265a/Fiji-Focus-Issue-08-Volume-10-April-24,-2019.aspx (Accessed: 21 January 2020).

Green Climate Fund (2017) *Mainstreaming gender in green climate fund projects*. Songdo. Available at: https://www.greenclimate.fund/sites/default/files/document/guidelines-gcf-toolkit-mainstreaming-gender_0.pdf (Accessed: 6 May 2021).

Green Climate Fund (2018) *GCF gender equality and social inclusion policy and action plan 2018–2020*. Songdo.

Green Climate Fund (2021) *Green climate fund projects & programmes*. Available at: https://www.greenclimate.fund/countries?f%5B0%5D=field_country%253Afield_group%3ASmall Island Developing States (Accessed: 13 May 2021).

Guadagno, E. (2016) 'Planned relocation : Lessons from Italy', *Migration, environment and climate change: Policy brief series*, 2(7). Available at: https://www.academia.edu/30068494/Planned_relocation_Lessons_from_ Italy (Accessed: 20 June 2019).

Haines, P. (2016) *Choiseul Bay township adaptation and relocation program, Choiseul Province, Solomon Islands*. Gold Coast.

Hermann, E. and Kempf, W. (2017) 'Climate change and the imagining of migration: Emerging discourses on Kiribati's land purchase in Fiji', *The Contemporary Pacific*, 29(2), pp. 231–263. doi: https://doi.org/10.1353/ cp.2017.0030.

Heslin, A. et al. (2019) 'Displacement and resettlement: understanding the role of climate change in contemporary migration', in Mechler, R. et al. (eds) *Loss and damage from climate change*. Springer, Cham, pp. 237–258. doi: https://doi.org/10.1007/978-3-319-72026-5_10.

Hino, M., Field, C. B. and Mach, K. J. (2017) 'Managed retreat as a response to natural hazard risk', *Nature Climate Change*, pp. 364–371. doi: https:// doi.org/10.1038/NCLIMATE3252.

Hirsch, E. (2015) '"It won't be any good to have democracy if we don't have a country": Climate change and the politics of synecdoche in the Maldives', *Global Environmental Change*, 35, pp. 190–198. doi: https://doi.org/10. 1016/j.gloenvcha.2015.09.008.

Hugo, G. (2011) 'Future demographic change and its interactions with migration and climate change', *Glob. Environ. Change*, 215, pp. 521–533.

IDMC (2021) *GRID 2021: Internal displacement in a changing climate*. Geneva. Available at: https://www.internal-displacement.org/sites/default/files/pub lications/documents/grid2021_idmc.pdf (Accessed: 26 July 2021).

International Labour Organization (2018) *Care work and care jobs for the future of decent work*. Geneva. Available at: https://www.ilo.org/wcmsp5/groups/ public/---dgreports/---dcomm/---publ/documents/publication/wcms_6 33135.pdf (Accessed: 30 July 2021).

Jalal, I. (2008) 'Legal protection against discrimination in Pacific Island countries', *The Equal Rights Review*, 2, pp. 32–39. Available at: https://www.equ alrightstrust.org/ertdocumentbank/Jalalarticle.pdf (Accessed: 22 September 2020).

Jolliffe, J. (2016) *Narikoso relocation project cost-benefit analysis update note*.

Kabeer, N. (2015) 'Gender, poverty, and inequality: A brief history of feminist contributions in the field of international development', *Gender and Development*, 23(2), pp. 189–205. doi: https://doi.org/10.1080/13552074.2015. 1062300.

Kate, T. (2021, February 18) Ali calls on PM Bainimarama to retract statement against Dr Chand. *The Fiji Times*. Available at: https://www.fijitimes.com/ ali-calls-on-pm-bainimarama-to-retract-statement-againstdr-chand/.

Kopf, A., Fink, M. and Weber, E. (2020) 'Gender vulnerability to climate change and natural hazards', in Amin, S. N., Watson, D., and Girard, C. (eds) *Mapping security in the Pacific*. London: Routledge, pp. 119–132. doi: https://doi.org/10.4324/9780429031816-12.

Kothari, U. and Arnall, A. (2019) 'Everyday life and environmental change', *The Geographical Journal*, pp. 130–141. doi: https://doi.org/10.1111/geoj.12296.

Kronsell, A. (2018) 'WPS and climate change', in Davies, S. E. and True, J. (eds) *The Oxford handbook of women, peace, and security*. Oxford University Press, pp. 726–736. doi: https://doi.org/10.1093/oxfordhb/9780190638276.013.55.

Lee-Koo, K. (2018) 'The gendered state and the emergence of a postconflict, postdisaster, semiautonomous state ', in Parashar, S., Tickner, J. A., and True, J. (eds) *Revisiting gendered states: Feminist imaginings of the state in international relations*. New York: Oxford University Press. doi: https://doi.org/10.1093/oso/9780190644031.001.0001.

Lewis, N. A. et al. (2020) 'Using qualitative approaches to improve quantitative inferences in environmental psychology', *MethodsX*, p. 100943. doi: https://doi.org/10.1016/j.mex.2020.100943.

Luetz, J. and Havea, P. H. (2018) '"We're not refugees, we'll stay here until we die!"—Climate change adaptation and migration experiences gathered from the Tulun and Nissan Atolls of Bougainville, Papua New Guinea', in *Climate change management*. Springer, pp. 3–29. doi: https://doi.org/10.1007/978-3-319-70703-7_1.

Marino, E. (2012) 'The long history of environmental migration: Assessing vulnerability construction and obstacles to successful relocation in Shishmaref, Alaska', *Global Environmental Change*, 22(2), pp. 374–381. doi: https://doi.org/10.1016/j.gloenvcha.2011.09.016.

Mavisakalyan, A. and Tarverdi, Y. (2019) 'Gender and climate change: Do female parliamentarians make difference?', *European Journal of Political Economy*, 56, pp. 151–164. doi: https://doi.org/10.1016/j.ejpoleco.2018.08.001.

McAdam, J. (2010) *Climate change and displacement : Multidisciplinary perspectives, climate change and displacement: Multidisciplinary perspectives*, in J. McAdam (ed). London: Hart Publishing. doi: https://doi.org/10.5040/9781472565211.

McAdam, J. (2013) 'Conceptualising climate change related movement', in *Climate change, forced migration, and international law*, pp. 583–605. doi: https://doi.org/10.1093/acprof.

McAdam, J. (2014) 'Historical cross-border relocations in the pacific: Lessons for planned relocations in the context of climate change', *Journal of Pacific History*. Routledge, pp. 301–327. doi: https://doi.org/10.1080/00223344.2014.953317.

McDonald, M. (2012) 'Human security and the politics of security', in Altman, D., Camilleri, J. A., and Hoffstaedter, G. (eds) *Why human security matters: Rethinking Australian foreign policy.* Sydney: Taylor & Francis Group, pp. 107–126. doi: https://doi.org/10.1080/14678802.2011.572461.

McLeman, R. and Gemenne, F. (2018) *Routledge handbook of environmental displacement and migration.* 1st edn, in McLeman, R. and Gemenne, F. (eds). London: Routledge. doi: https://doi.org/10.4324/9781315638843.

Meo-Sewabu, L. (2016) 'Na Marama iTaukei Kei Na Vanua: Culturally embedded agency of indigenous Fijian women-opportunities and constraints', *New Zealand Sociology*, 31(2), pp. 96–122.

Ministry for Social Welfare, W. & P. A. (2014) *Fiji national gender policy.* Suva. Available at: https://www.fiji.gov.fj/getattachment/db294b55-f2ca-4d44-bc81-f832e73cab6c/NATIONAL-GENDER-POLICY-AWARENESS.aspx (Accessed: 16 June 2021).

Ministry of Economy (2019) *National climate change policy 2018–2020.* Suva. Available at: https://www.economy.gov.fj/images/CCIC/uploads/General/FIJI-National-Climate-Change-Policy-2018-2030-FINAL.pdf (Accessed: 16 June 2021).

Ministry of Economy (2020) *Fiji climate finance snapshot 2016–2019.* Suva. Available at: https://www.economy.gov.fj/images/CCIC/uploads/ClimateFinance/Fiji-Climate-Finance-Snapshot-2016-2019.pdf (Accessed: 1 October 2021).

Ministry of Economy, F. I. (2018) *Planned relocation guidelines a framework to undertake climate change related relocation.*

Ministry of Economy, F. I. (2019) *Displacement guidelines in the context of climate change and disasters.*

Ministry of Foreign Affairs and International Cooperation, F. I. (2014) *Republic of Fiji second national communication to the united nations framework convention on climate change.* Suva.

Nabobo-Baba, U. (2008) 'Decolonising framings in Pacific research: Indigenous Fijian Vanua research framework as an organic response', *Alternative: An International Journal of Indigenous Peoples*, 4(2), pp. 140–154. Available at: https://journals-sagepub-com.ezproxy.lib.monash.edu.au/doi/pdf/10.1177/117718010800400210 (Accessed: 30 September 2019).

Ourbak, T. and Magnan, A. K. (2018) The Paris agreement and climate change negotiations: Small Islands, big players. *Reg Environ Change*, 18, pp. 2201–2207. doi: https://doi.org/10.1007/s10113-017-1247-9.

Pacific Islands Forum (2019) *Kainaki II declaration for urgent climate action now—Forum Sec.* Available at: https://www.forumsec.org/2020/11/11/kainaki/ (Accessed: 5 May 2021).

Pascoe, S. (2015) 'Sailing the waves on our own: Climate change migration, self-determination and the carteret islands', *QUT Law Review*, 15(2), pp. 72–85. Available at: https://search-informit-com-au.ezproxy.lib.monash.edu.au/documentSummary;res=IELHSS;dn=891927198137587 (Accessed: 20 June 2019).

Piggott-McKellar, A. E., Pearson, J., et al. (2019a) 'A livelihood analysis of resettlement outcomes: Lessons for climate-induced relocations', *Ambio*. Springer. doi: https://doi.org/10.1007/s13280-019-01289-5.

Piggott-McKellar, A. E., McNamara, K. E., et al. (2019b) 'Moving people in a changing climate: Lessons from two case studies in Fiji', *Social Sciences*, 8(5). doi: https://doi.org/10.3390/socsci8050133.

Platform on Disaster Displacement (2021) 'Fiji assumes chairmanship of the platform on disaster displacement—Disaster displacement'. Available at: https://disasterdisplacement.org/fiji-assumes-chairmanship-of-the-platform-on-disaster-displacement (Accessed: 16 June 2021).

Podesta, J. (2019) *The climate crisis, migration, and refugees*, *Brookings*. Available at: https://www.brookings.edu/research/the-climate-crisis-migration-and-refugees/ (Accessed: 14 April 2020).

Ratuva, S. and Lawson, S. (2016) *The people have spoken: The 2014 elections in Fiji*, *The people have spoken: The 2014 elections in Fiji*, in Ratuva, S. and Lawson, S. (eds). ANU Press. Available at: http://press.anu.edu.au/publications/series/pacific-series/people-have-spoken (Accessed: 4 October 2019).

Republic of Fiji (2021) *Budget estimates 2021–2022*. Suva. Available at: https://www.economy.gov.fj/images/Budget/budgetdocuments/estimates/BUDGET_ESTIMATES_2021-2022_Web.pdf (Accessed: 4 August 2021).

Rigaud, K. K. *et al.* (2018) *Groundswell: Preparing for internal climate migration*. Washington D.C. Available at: https://openknowledge.worldbank.org/handle/10986/29461.

Rika, N. (2018, July 11) Bainimarama accused of abusing opposition leader. *Islands Business*. Available at: https://www.islandsbusiness.com/past-news-break-articles/item/2149-bainimarama-accused-of-abusing-oppositionleader.html.

Samuwai, J. and Fihaki, E. (2020) *Making climate finance work for women: Voices from Polynesian and Micronesian communities*. Suva. Available at: https://www.pasifikarising.org/making-climate-finance-work-for-women/ (Accessed: 15 May 2020).

Seymour, E. (2020) *Climate change in women, peace and security national action plans*. Available at: https://www.sipri.org/sites/default/files/2020-06/siprinsight2007.pdf (Accessed: 16 November 2020).

Sheldrick, D. (2016) 'Move to Iceland: Fiji PM's advice for gay couples', *SBS*, 7 January. Available at: https://www.sbs.com.au/topics/pride/article/

2016/01/07/move-iceland-fiji-pms-advice-gay-couples (Accessed: 27 August 2021).

Simpson, S. (2020) 'Lessons in climate-driven relocation: The Narikoso case | Earth Journalism Network'. MaiTV. Available at: https://earthjournalism. net/stories/lessons-in-climate-driven-relocation-the-narikoso-case (Accessed: 4 August 2021).

Singh, P. et al. (2020) 'Place Attachment and cultural barriers to climate change induced relocation: Lessons from Vunisavisavi village, Vanua Levu, Fiji', in *Climate change management*. Springer, pp. 27–43. doi: https://doi.org/10. 1007/978-3-030-40552-6_2.

Tanyag, M. and True, J. (2019) 'Gender-responsive alternatives on climate change from a feminist standpoint', in *Climate hazards, disasters, and gender ramifications*. Routledge, pp. 29–47. doi: https://doi.org/10.4324/978042 9424861-2.

The Nansen Initiative (2015) *The Nansen initiative: agenda for the protection of cross-border displaced persons in the context of disasters and climate change volume I*. Available at: https://disasterdisplacement.org/wp-content/upl oads/2014/08/EN_Protection_Agenda_Volume_I_-low_res.pdf (Accessed: 16 June 2020).

The World Bank (2017) *Climate change and disaster management—Pacific possible background paper No. 6*. Available at: https://reliefweb.int/report/ world/climate-change-and-disaster-management-pacific-possible-background-paper-no6 (Accessed: 6 February 2020).

Thornton, F. et al. (2020) 'Climate crisis and local communities multiple mobilities in Pacific Islands communities', *Forced Migration Review*, (64), pp. 32–35. Available at: https://www.fmreview.org/sites/fmr/files/FMRdow nloads/en/issue64/Pacific-mobilities.pdf (Accessed: 22 July 2020).

True, J. (2012) 'What has poverty got to do with it? ', in *The political economy of violence against women*. New York: Oxford University Press. doi: https:// doi.org/10.1093/acprof:oso/9780199755929.001.0001.

True, J. (2016) *Women, peace and security in Asia Pacific: Emerging issues in national action plans for women, peace and security, Asia-Pacific regional symposium on national action plans on women*. Available at: http://www2.unwomen.org/-/media/fieldofficeeseasia/docs/pub lications/2016/12/1-nap-jt-for-online-r4.pdf?la=en&vs=2213 (Accessed: 20 June 2019).

UN Environment Programme (2020) *Frank Bainimarama—Policy leadership award | Champions of the earth*. Available at: https://www.unep.org/champi onsofearth/laureates/2020/frank-bainimarama (Accessed: 5 May 2021).

UN Women (2014) *Climate change, disasters and gender-based violence in the Pacific*.

UNFCCC (2002) *Part two: Action taken by the conference of parties.* Marrakech. Available at: https://unfccc.int/sites/default/files/resource/docs/cop7/13a04.pdf (Accessed: 10 June 2021).

UNFCCC (2013) *Warsaw international mechanism for loss and damage | UNFCCC.* doi: https://doi.org/10.1126/science.202.4366.409.

UNFCCC (2019a) 'Gender and climate change', in *Conference of parties twenty-fifth session.* Madrid: UNFCCC. Available at: https://unfccc.int/sites/default/files/resource/cp2019a_L03E.pdf (Accessed: 29 April 2021).

UNFCCC (2019b) *Warsaw International mechanism for loss and damage associated with climate change impacts and its 2019b review.* Available at: https://unfccc.int/sites/default/files/resource/cma2_auv_6_WIM.pdf (Accessed: 10 June 2021).

UNFCCC (2021) *Overview of outputs from the Warsaw international mechanism for loss and damage (WIM) and its executive committee | UNFCCC.* Available at: https://unfccc.int/process-and-meetings/bodies/constituted-bodies/executive-committee-of-the-warsaw-international-mechanism-for-loss-and-damage-wim-excom/workshops-meetings/2019-wim-review-event/overview-of-outputs-from-the-warsaw-international-mechanism-for-loss-and#eq-6 (Accessed: 10 June 2021).

UNFCCC United Nations Framework Convention on Climate Change (2010) *Report of the conference of the parties on its sixteenth session, held in Cancun from 29 November to 10 December 2010, conference of the parties on its sixteenth session.* Available at: https://unfccc.int/resource/docs/2010/cop16/eng/07a01.pdf (Accessed: 20 June 2019).

United Nations (2015) *Paris agreement.* https://unfccc.int/sites/default/files/english_paris_agreement.pdf.

United Nations (2018) *Global compact for safe, orderly and regular migration.* https://refugeesmigrants.un.org/sites/default/files/180713_agreed_outcome_global_compact_for_migration.pdf.

United Nations (2019) 'UN chief lauds Fijians as "natural global leaders" on climate, environment, hails "symbiotic relationship" with land and sea | | UN News', *UN News*, 16 May. Available at: https://news.un.org/en/story/2019/05/1038581 (Accessed: 17 August 2020).

United Nations (2020) *Human rights committee: Views adopted by the committee under article 5 (4) of the optional protocol, concerning communication No. 2728/2016.* Available at: https://tbinternet.ohchr.org/_layouts/15/treatybodyexternal/Download.aspx?symbolno=CCPR%2FC%2F127%2FD%2F2728%2F2016&Lang=en (Accessed: 17 June 2020).

United Nations Office for Disaster Risk Reduction (2015) *Sendai framework for disaster risk reduction 2015–2030.* Sendai. Available at: https://www.preventionweb.net/files/43291_sendaiframeworkfordrren.pdf (Accessed: 16 June 2020).

Vakasukawaqa, A. (2019, September 6) Tabuya issue thrown out of Parliament. *The Fiji Times*. Available at: https://www.fijitimes.com/tabuya-issue-thrown-out-of-parliament/?fbclid=IwAR2tcl1mj23ROamgr_FfPWBXqa9sctkbhat_l OI3M3Z8qUa3SeIQZ11N1HY.

Vitukawalu, B. et al. (2015) *Addressing barriers and constraints to gender equality and social inclusion of women seafood sellers in municipal markets in Fiji*. Available at: https://unwomen.org.au/newsroom/ (Accessed: 16 June 2020).

Voigt-Graf, C. and Kagan, S. (2017) 'Migration and labour mobility from Kiribati', *SSRN Electronic Journal*. doi: https://doi.org/10.2139/ssrn.293 7416.

Wasuka, E. (2020) 'Landmark decision from UN human rights committee paves way for climate refugees—ABC News', *ABC News*, 21 January. Available at: https://www.abc.net.au/news/2020-01-21/un-human-rights-ruling-wor lds-first-climate-refugee-kiribati/11887070 (Accessed: 27 January 2021).

Weber, E. (2015) 'Envisioning South–South relations in the fields of environmental change and migration in the Pacific Islands—Past, present and futures', *Bandung: Journal of the Global South*, 2(1). doi: https://doi.org/10.1186/s40728-014-0009-z.

WEDO, Women's Environment and Development Organisation (2020) *UNFCCC: Progress on achieving gender balance*. Available at: https://wedo.org/wp-content/uploads/2020/01/Factsheet-UNFCCC-Progress-Achiev ing-Gender-Balance-2019.pdf (Accessed: 2 July 2020).

World Bank (2017) *Climate vulnerability assessment: Making Fiji climate resilient*. Available at: http://documents.worldbank.org/curated/en/163 081509454340771/Climate-vulnerability-assessment-making-Fiji-climate-res ilient (Accessed: 4 October 2019).

Assertion of Indigenous Identity in the Face of Climate Change: The Works of Two Millennial Paiwan Authors

Fanny Caron

Typhoon Morakot, the deadliest in Taiwan's recorded history, hit the island on the 8th of August 2009. Its severity, linked to anthropogenic climate change[1] caused the destruction of Paiwan 排灣[2] villages, leading to the relocation of its inhabitants—a new hardship for people with a centuries-old colonial history that implied multiple forced displacements. Nonetheless, whilst warning about the catastrophic consequences of dominant societies' disrespect for nature, budding Paiwan authors Tjinuay Ljivangerau (Chen Mengjun 陳孟君, 1983–), in her poem [Moving. Formosa] ("Yidong. Fuermosha" 「移動·福爾摩沙」), and Ising Suaiyung (Zhu Keyuan 朱克遠), with his dystopian short story [Crimson Earth] ("Chitu" 「赤土」), used this climatic phenomenon to

---

[1] See, for example: Wang et al. (2019: 3454–3464).

[2] The Paiwan are the second largest of sixteen officially recognised Indigenous Peoples of Taiwan, numbering 102 730 (0.44% of Taiwan's predominantly Han population), as of January 2020: https://www.cip.gov.tw/portal/docDetail.html?CID=B68D98A9742D94C7&DID=0C3331F0EBD318C2087B2099BEBAC7E8 (accessed: 15 May 2021).

F. Caron (✉)
IrAsia, Aix-Marseille University-CNRS, Marseille, France

N. J. P. Alsford (ed.), *Pacific Voices and Climate Change*,
https://doi.org/10.1007/978-3-030-98460-1_6

assert their Indigeneity and resilience, and to create networks of kinship with their international Indigenous[3] families.

By being particularly attentive to these Paiwan authors' tales, readers everywhere can take heed of their admonition and be made aware of Indigenous Peoples' ancestral scientific knowledge. Therefore, this paper aims to demonstrate how Ljivangerau and Suaiyung, heirs of Paiwan storytellers, by focusing on an event linked to climate change, contribute to a literature that opens a path to tribal imaginary, and thus Indigenous identity. Indigenous scholars and authors aspire to focus on the (re)writing of their own (hi)stories and culture as well, by building "a historical consciousness based on Indigenous subjectivity" (Sun 2009: 10). With an approach to literary and cultural analysis drawn from an emic perspective, I strive to organise a dialogue between contemporary Paiwan authors in order to highlight how they perpetuate their tradition of resilience (the capacity to promptly recover from a traumatic ordeal). This approach makes it possible to rethink the place of Paiwan works on the world's literary scene. It offers a new outlook on literary criticism, having as its source the studied culture itself, and it would thus enable us to develop new analysis models of tribal societies.

Moreover, as alliances between Indigenous Peoples from far and wide are ever growing, Paiwan authors are actively participating in international meetings, where they make their voices heard on a globalised political, cultural, and literary scene. Considering our current global ecological and climatic situation, Paiwan authors' ethics of resilience and *survivance* can be an alternative organising model for non-Indigenous Peoples who will increasingly suffer from climate change. This notion of survivance, a *portmanteau* word (from "survival" and "resistance"), used by Anishinaabe cultural theorist Gerald Vizenor, transcends merely surviving or reacting to struggles. It implies "the continuation of Indigenous stories", even when told in the colonisers' language. Hence, "Native survivance stories are renunciations of dominance, tragedy and victimry" (1999: vii).

With the first waves of Dutch (1624–1662), Zheng 鄭 (1662–1683), and Qing 清 (1683–1895) colonisations, the Paiwan witnessed their land

---

[3] In this paper, "Indigenous", "Native", and even "People(s)" are capitalised when referring to the original inhabitants of a place, in accordance with guidelines on terminology and capitalisation of Indigenous Peoples themselves, who chose to redress "mainstream society's history of regarding Indigenous Peoples as having no legitimate national identities" (Younging 2018: 77).

being invaded and pillaged, with thousands of trees being cut down for economic benefits and the construction of infrastructures. Thereupon, the arrival of two consecutive imperialist blocks, the Japanese (1895–1945) and the Kuomintang (1945–),[4] both imposing their own value systems, had a violent impact on the local culture and identity. However, in precolonial times, the Paiwan, who inhabited the southern part of Taiwan's Central Mountain Range, lived in a society based on oral tradition. Each village's history and stories were passed on from generation to generation, allowing its members to develop a strong sense of tribal belonging. To cope with settlers' acculturation attempts, they adopted resistance strategies to not only preserve, but also adapt, their stories into a contemporary literature, born in the midst of the 1980s Indigenous Movements (*Yuanzhumin yundong* 原住民運動), where the island's First Inhabitants came together to fight for their rights and for their agency.

If "Paiwaness" is intrinsically bound to an oral tradition detailing the People's relationship with their ecosystem, what are these stories' deeper meaning? And how do they translate into a contemporary literature of resilience? In addition, how were environmental issues brought up by the first generations of authors recaptured by millennial ones by means of an event manifesting the implications of climate change—Typhoon Morakot? Founded on tribal societies' organic model, rooted in a natural space devoid of technical dominance and economical exploitation, Taiwan Indigenous literature is committed to a cause: the defence of core ancestral values and knowledge, linked to the land and all its inhabitants. It became a vector through which Indigenous authors reclaim their culture, (hi)stories, and territories. They likewise reclaim some of the settlers' jargon, including the term "tribe" (*buluo* 部落), recurrent in their works, and used to describe the social group composed of villagers interconnected via intricate economical, religious, cultural, and kinship webs. Rukai (*Lukai zu* 魯凱族) anthropologist Sasala Taiban 撒沙勒·台邦 (1965–), who advocated for the concept of tribalism (*buluo zhuyi* 部

---

[4] The Chinese Nationalist Party, or Kuomintang (*Zhongguo guomindang* 中國國民黨), founded in Mainland China in 1912 by Sun-Yat-sen, was the governing party in Taiwan after Japan's retrocession of the island in 1945, and the only authorised political party until 1986. For Indigenous People, this Han colonial rule never truly ended, since their motherland is still occupied, and they never (re)gained their right to self-determination or to tribal sovereignty.

落主義) as a tool for the "rebuilding of traditional society",[5] applies it as a fundamental axiom of the Indigenous Movements (1993: 28–40; 2006).

This paper will start with a brief overview of flood stories and excerpts from early Paiwan literary works providing environmental lessons and warnings. Established authors like Malieyafusi Monaneng 馬列雅弗·莫那能 (1956–), the first Indigenous writer to publish a collection of poems in Taiwan (1989), Liglav Awu 利格拉樂·啊(女烏) (1969–) and Ahronglong Sakinu 亞榮隆·撒可(1972–), whose early works containing life stories and tribal histories were published in the 1990s, denounce the hegemonic pillaging of tribal land and culture. All paved the way for Ljivangerau and Suaiyung's literary productions, emblematic of Indigenous Peoples' concerns regarding our "modern" way of life and consumerism. Hence, the remainder of this paper shall focus on Suaiyung's storytelling as a form of ecological plea and on Ljivangerau's use of typhoons imagery to "weave" a poetic multicultural tapestry. Finally, I will illustrate how these authors offer readers, worldwide, life sciences "tools" inherited from their forebears' alternate ways of interacting with nature.

Environmental Lessons and Warnings: From Oral Stories to an Indigenous literature's Advent

Indigenous ancestral stories, in particular cosmogonic ones, "contained historic implications" (Poiconu 2012: 271) providing answers to onto-logical questions and bringing to light People's attachment to the land and all its denizens.

Flood Stories: The Embodiment of Precolonial Eco-Conscience

In Paiwan deluge narratives,[6] many creatures contribute to human survival, and even earthworms participate in the (re)creation of the world.

---

[5] All translations from Mandarin to English in this paper are by the author.

[6] These stories were collected in Paiwan by Japanese linguists Ogawa Naoyoshi 小川尚義 and Asai Erin 淺井惠倫 in 1932, then translated in English by John Whitehorn in *One Hundred Paiwan Texts* (2003); and by Japanese anthropologists Kojima Yudo 小島由道 and Kobayashi Yasuyoshi 小林保祥 for the [Investigation Report on Barbarian tribes' Customs, Volume 5: The Paiwan, Book 1]《番族慣習調查報告書第五卷. 排灣族第一冊》published in Japanese in 1920, then edited by Chiang Bien 蔣斌 and published in Mandarin by Taiwan's Academia Sinica in 2003.

Their excrements become the "soil covering the ground", turning a devastated landscape with "no solid ground left" into fertile farmland (Early and Whitehorn 2003: 330–302; 236–240), and bringing mountains to a world once made only of plains (Chiang [1920] 2003: 121–123). This indicates that in precolonial times Paiwan People already understood that their physical survival hinged on the ecosystem's fine balance, where seemingly insignificant creatures like earthworms played a crucial role in soil fertility. A fact our non-Indigenous societies are only now grasping, when intensive and repetitive use of pesticides and fertilisers rendered many soils infertile, with a detrimental effect on the surrounding fauna and flora (Bourguignon and Bourguignon 2015: 49–50; Jeunehomme 2018: 3), as well as on the climate (Jet Propulsion Laboratory/NASA 2008).

Furthermore, these stories convey the criticality of trees and vegetation. For instance, the only remaining couple survives by holding on to tallgrass, a plant capable of withstanding the storm. And after the torrential rains, there were "landslides everywhere", until earthworms excreted the soil where plants started to grow (Early and Whitehorn 2003: 236–240). Another version of the story states that the deluge was a punishment from the spirits, after Paiwan forefathers violated a rule prohibiting them from entering a sacred forest to cut down some trees (Chiang [1920] 2003: 122). These elements could very well reveal a Paiwan precolonial knowledge of the link between plants and trees and stronger soils. Today, deforestation is listed as one of the main causes of climate change (Streiff 2021) and is part of a vicious cycle, as seen during Typhoon Morakot: climate change—in part due to deforestation—caused extreme rainfall (Wang et al. 2019: 3459–3460; Chen 2016); and increased soil erosion—also due to deforestation (Jeunehomme 2018: 6), coupled with said rainfall, led to disastrous hillslopes landslides on Paiwan and Rukai territories (Chen 2016).

Paiwan stories, based on a thorough analysis and knowledge of the natural world, were considered mere folktales by settlers. Japanese anthropologists and linguists who collected them in the first half of the twentieth century were more interested in their linguistic than in their cultural or didactic content. At a time when Japanese policemen were incidentally in charge of Paiwan children's education, the Japanese value system, focused on "progress" and industrialisation, was set against—and incompatible with—the Paiwan system, tuned to nature conservation. But the People endured, and so did their stories, like flood narratives, emblematic of

Paiwan "ecological"[7] values. After being passed down over generations, they were reinterpreted in a literature of resilience.

The Voices of Established Paiwan Authors

With the arrival of the Kuomintang in the 1940s, whose rule was perhaps even worse than its predecessor's (Simon 2002), the Paiwan were hit by yet another harrowing wave of colonisation. Under the Kuomintang's centralised educational system, Indigenous children were forced to identify to the Chinese nation by learning its history, geography, culture, and value system, and by only speaking Mandarin. At school, where the first generations of Paiwan writers and scholars were "educated", these children were called "monkeys", "barbarians", and "children of the mountain people" by Han teachers and schoolmates alike (Tung 1995: 7; Awu 1996: 34, 36–37).

Paiwan authors, such as Monaneng, from the village of Aljungic (*Anshuo cun* 安朔村), thus created a literature of survivance and resilience in which they shared their experience and (hi)stories. Monaneng, who spoke out on his People's issues during the Indigenous Movements, became an actant of Paiwan resilience. In his collection of poems, [Beautiful Ears of Rice] *(Meili de daosui*《美麗的稻穗》), he gave a voice to the victims of colonisation and called out for them to overcome their tragic past, encouraging, as Vizenor put it, the "continuation" of their stories (1999: vii).

In his poem [Burning] (*Ranshao*「燃燒」), he refutes the Kuomintang's assertion that it liberated Taiwan from the Japanese, since it perpetuated and intensified its predecessors' practices of territorial appropriation and acculturation ([1989] 2014: 055–056):

中國你來了,	China, you came,
[...]	[...]
從日本人手中接掌,	From the Japanese's hands you took,
所有的錢勢和財產,	All the money and properties,
然後你對我說:	Then you told me:
[...]	[...]
不能打獵,	Hunting is forbidden,

---

[7] The term "ecological" is deliberately used here as an anachronical concept allowing us to define Paiwan precolonial ethics of relations between the People and the(ir) environment—fauna and flora.

不能講母親的話語,	Speaking in our mother tongue is forbidden,
……	…
什麼民族、民權、民生的,	You say the Three Principles of the People,
說是為人們實施的制度。	Implement a system for the People
但我卻,	But for me, however,
失去的愈來愈多。	The losses are increasingly numerous
土地?	The land?
野獸?	Wild animals?
自由?	Liberty?
自尊!	Dignity!

This change in tone, from questioning what is lost to demanding dignity for all Indigenous Peoples, demonstrates Monaneng's survivance spirit, his refusal to be subjugated. In fact, in another poem, he instructs his People, who are "no longer weak", not to be "used" as instruments of the government's "authoritarian domination" (*Ibid.* [1989] 2014: 076). Monaneng's poetry highlights the settlers' many offences, including their violation of the island's lush environment. In [Answer] (*Huida* 「回答」), he mentions "throats suffocated and obstructed by the city's spreading smog" (*Ibid.*, 081). Then, in [The hundred-pace snake is dead] (*Baibushe sile* 「百步蛇死了」), he draws a parallel between the grim fates, brought on by the Han, of a young Paiwan girl sold into prostitution and her mythical ancestor, the hundred-pace snake, killed and "packed in a large transparent medicine bottle" with a label that reads "aphrodisiac" (*Ibid.*, 170–171).

Likewise, Liglav Awu, the daughter of a Paiwan mother and a Chinese Mainlander father, voices her concerns for the descendants of the hundred-pace snake in 1990s Taiwan, when the dominant society's capitalist ambitions were once more leading it to exploit her People's natural resources. Awu's first book, [Who Will Wear the Beautiful Clothes I Weave] (*Shei lai chuan wo zhi de meili yishang* 《誰來穿我織的美麗衣裳》), contains stories describing her mother's life in a military dependents' village (*juancun* 眷村) where Indigenous women were perceived as "ferocious beasts" by the retired soldier's Mainlander wives (1996: 36), and her homecoming almost two decades after she left her native tribe. The momentous occasion of her mother's return would be tainted by an old threat in the guise of a new opportunity: the further deterioration, a few years later, of tribal land, peace, and community (1996: 39):

只是，這個部落又能庇佑百步蛇的子孫多久呢?我聽到怪手正在怒吼著，那是資本家正在部落外二公里處，為開發新的觀光資源而動工著，原本在山上工作的部落少年，為了較多的收入，放棄了祖先留下的小米田，紛紛投入開發的工作了，但卻不知道那正是部落的水源地啊!我不禁害怕，部落外這一批又一批的文明獸，如此蓄勢待發的準備一擁而上，這樣美麗與寧靜的部落，還能維持多久?

But how much longer will this tribe be able to protect the descendants of the hundred-pace snake? I hear the excavators roaring. It is the capitalists who started developing new touristic resources two kilometres away from the tribe. The young tribespeople who once worked on the mountains have abandoned the millet field left by the ancestors. One after another, they have thrown themselves in this development work for a comparatively higher income. Oh, but they do not know that the water sources of the tribe are precisely here! I cannot help but be afraid when I see those successive hordes of civilised beasts outside of the tribe. Therefore, we must prepare by promoting our strength through unity. Such a beautiful and peaceful tribe, how much longer can it be preserved?

To this popular portrayal of Paiwan People as "ferocious beasts" painted by the Han, Awu counters with a depiction of the latter as "civilised beasts". In her second book, [Red Mouthed *Vuvu*], (*Hong zuiba de VuVu*《紅嘴巴的VuVu》), Awu relates her visit to the village of Heping (*Heping cun* 和平村) where she witnessed the Truku People (*Tailuge zu* 太魯閣族) battling against the *Asia Cement Corporation* (*Yazhou shuini gufen youxian gongsi* 亞洲水泥股份有限公司), who were mining Truku's ancestral territories ([1997] 2001: 163–181). Canadian anthropologist and sociologist Scott Simon indicates that after being forcibly relocated from their homes in the mountains to the foothills by the Japanese, then "from the foothills to the base of the mountains" by the Kuomintang who forbade them to hunt or fish on their native soil, the Truku "are now forced to live on narrow strips of land between the mountains and the sea, amidst cement quarries, cement factories, railroad tracks, and industrial parks" (2002). The ongoing Truku's plight echoes all of Taiwan Indigenous Peoples' battles for their land. Additionally, this island-wide deforestation, in conjunction with major road construction and climate change (causing stronger typhoons), has led to an acceleration of landslide incidence (Chang and Slaymaker 2020).

Author Ahronglong Sakinu shares Awu's urge to return to tribal wisdom. Just as her inspiration comes from her Paiwan mother and grand-mother, his comes from his Paiwan hunter father and grandfather. In his first book [Boar. Flying Squirrel. Sakinu] (*Shanzhu. Feishu. Sakenu*《山豬. 飛鼠. 撒可努》), he states ([1998] 2011: 13–14):

> 開始動筆寫東西，最大的原動力來自於祖父智慧的催促，再者就是父親和山林奧妙、美麗、人格化的相處之道。小時候大自然是我學習的教室 [...]。在那裡我得到族人世代教授的智慧，也看到了排灣族人用生命在自然世界裡累積的經驗。

> When I began writing, my greatest motivation came from my grandfather's wisdom urgings, as well as from the profound, beautiful and personal way my father got along with the mountain forests. When I was a child, nature was the classroom in which I studied. [...] It is there that I received the wisdom taught for generations by my tribesmen, and where I saw how the Paiwan People accumulated experience of the natural world at the cost of their own lives.

At a time when very few elements of his Native culture had been preserved in his home tribe of Lalaulan (*Laolaolan buluo*, 牢勞蘭部落), Sakinu advocated for the resurgence of the ancestral ways, and left Taipei to return to Lalaulan in the early 2000s. To share his forefathers' wisdom, he recorded their stories in his books, and established a Hunter School in Lalaulan in 2005, where he relays Paiwan hunters' empathetic approach towards animals, made evident in his father's instructions to "Treat animals as you would humans, consider yourself as an animal, and you will then understand their behaviour and their language" ([1998] 2011: 29).

Like Monaneng and Awu, Sakinu questions dominant societies' perspective on civilisation. During a hunting trip with his father, where he learned how boars use their snout to find buried bamboo sprouts, Sakinu's father pointed out that, as opposed to boars' "civilisation", "human civilisation and science" can be "backwards, obsolete" since "science and technology aren't advanced enough to invent machines to find bamboo sprouts underground" ([2002] 2011: 67–68). Following profit and productivity ideologies, dominant societies conveniently view(ed) animals as insensitive creatures, and therefore created intensive stockbreeding methods that led to deforestation and climate change (Jet Propulsion Laboratory/NASA 2008). Paiwan authors contrast two civilisation models, one built on an ecologically sustainable value chain, and a non-Indigenous one, whose objective is, as Sakinu phrases it "development, control, substitution" ([2002] 2011: 68). This "ultra-liberal" civilisation on the verge of collapse, and the climatic devastation it brings with it, is a core concern for millennial Paiwan writers who built a narrative around Typhoon Morakot.

Suaiyung's Ecological Plea in [Crimson Earth]

Ising Suaiyung, from the Tjalja'avus tribe (*Laiyi buluo* 來義部落), is one such author. His short story [Crimson Earth] (*Chitu*「赤土」), set in a dystopian future, on a sixteen square kilometres islet named Crimson Earth, whose surface was rendered sterile and inhabitable by the leaders of a hegemonic power, is evidently connected to current environmental issues. In fact, Suaiyung, who describes the colour of this land as "brownish red" (Suaiyung 2010: 89), didn't use the generic term for red, *hong* 紅, in his title. He instead used the character *chi* 赤, referring to a kind of red, slightly lighter than vermillion, that can otherwise mean "bare" or "empty".

A Tale on the Ramifications of Climate Change

Decades before the story's main plot, the people living on Crimson Earth split into two societies: one remained on the surface, and one founded White Ring (*Bai huanquan* 白環圈), a giant subterranean glass tube world with "no soil, no sunlight" and "biotechnological" food grown from "marine microorganisms" (*Ibid.*, 90). Suaiyung's tale thus engages us to think about the impacts of wastefulness, the depletion of natural resources, and rapid urbanisation. It opens with a child remarking that (*Ibid.*, 89):

「爸爸，人是不能在這樣的試管內存活得，就算環境再優美，物算再豐富，經濟再富裕。你難道忘了小時候打水漂也是一種幸福嗎?人，是沒有辦法脫離土地生活的。」

Dad, in spite of the surrounding's beauty, in spite of the resources' richness and economy's prosperity, man cannot live in such a test tube. Can you have forgotten that making ricochets on water as a child is also a form of happiness? Man cannot live separated from the land.

Dong-Qing 東清, the hero/narrator, imprisoned in a small and dark cell ten metres underground, then proceeds to chronicle the aftermath of a massive explosion on Crimson Earth that happened two years prior.

Referred to as the "incident" (*shijian* 事件), it was neither a "natural disaster" nor an attack from a "neighbouring country", but "a kill order issued by White Ring" (*ibid.*, 96). A secret Dong-Qing uncovered, leading to his incarceration. The islet's surface is now "a cemetery" with "no traces of animal or plant life". Crimson Earth's survivors, forced to seek shelter from White Ring, were set up in a restricted area called Refuge (*Binansuo* 避難所), and live "without any partition or privacy" in an overcrowded "wobbly circus tent" (*Ibid.*, 93–94). White Ringers—perhaps a microcosm of Taiwan, and the world—are separated according to social, political, and economic status into different living areas, literal strata. Crimson Earthers live in the area closest to the toxic surface and are prohibited from entering lower areas, becoming White Ring's second-rate and poorest citizens, whilst the rich and powerful reside in the deepest part of the tube (*Ibid.*, 89–90).

The islet itself is reminiscent of the fifteen square kilometres Green Island (*Lu Dao* 綠島), originally inhabited by the Amis People (*Amei zu* 阿美族). The Amis saw their home transformed into a penal settlement for political prisoners during Taiwan's martial law era (1949–1987), just as Crimson Earthers suffered the actions of hegemonic leaders. Moreover, Suaiyung's narrative is abound with references to the Paiwan People, linking them to Crimson Earthers. A main character named Xiao-Ling 小凌 was born on Crimson Earth and adopted by White Ringers when she was a baby. Nonetheless, her "darker complexion" sets her apart from her adoptive countrymen who alienate, ostracise, and discriminate against her (*ibid.*, 95). As a child, she endured insults from her White Ring schoolmates, who told her Crimson Earthers were "barbarians" and called her a "lizards' daughter" (*Ibid.*, 91)—words echoing Awu's personal experience.

Gu-Jiang 古將, the "village head" (*cunzhang* 村長)—akin to a *mamazangiljan* (Paiwan "chief")—of Refuge, explains that White Ring officials, in order to "seize Crimson Earth's natural resources", committed "genocide" against his People. They ended up destroying everything else in the process, possibly accidentally, and then pretended to help the survivors by locking them up in "a cage" "worse than an animal's den" (*Ibid.*, 95–96). Again, this echoes another Paiwan author's words, in this case Monaneng's denunciation of the Kuomintang's actions.

Dong-Qing prisoner number is "2009–88", alluding to when Typhoon Morakot hit Taiwan (8 August 2009), leaving a trail of destruction amongst Paiwan and Rukai communities. As members of these communities were ordered to relocate by the Taiwanese government—raising past traumas of forced displacements—so were the Crimson Earthers by White Ring's government. Suaiyung is connecting his hero to a meteorological phenomenon whose severity, much like the events on Crimson Earth, was not brought on by nature but by men. The author hints even more overtly at anthropogenic climate change with Xiao-Ling's mention of Crimson Earth's "awful mess of a weather" brought on by White Ring's rulers, and the narrator's description of a "furiously whizzing wind", "countless huge waves" coming from the "black sea" and "red lighting" shooting from the sky (*Ibid.*, 99), evocative of Morakot's "unnatural" fierceness.

Furthermore, being a White Ring prisoner does not define who the hero is. Instead, his identity is linked with a natural element when he is told he was "born of bamboos" (*Ibid.*, 92), in reference to a Paiwan cosmogonic story according to which the People emerged from bamboo stems. A significant revelation that will trigger the hero's return to his ancestors' land.

A Return to the Ancestral Homeland

Early in the story, we are introduced to "strange old man" in an equally peculiar way—via sounds of his shovelling and ploughing the soil just above Dong-Qing's cell. Every day from hours on end, he is seen desperately trying to maintain ties with the earth, by digging and watering Crimson Earth's barren ground. This seemingly nonsensical conduct is, on the contrary, quite meaningful and perpetuates an age-old skill, safeguarding it from oblivion, despite its unsuitability on an apparently dead soil or in White Ring's artificial world.

According to Dong-Qing, the old man has been behaving oddly for decades, often muttering to himself or to passers-by "strange words": "We are the People of the bamboo grove, and we must return to the bamboo grove, to our ancestral land" (*Ibid.*, 90). Indeed, during the only discussion they had before the "incident", the Elder urged Dong-Qing to return to this "mythical" homeland (*Ibid.*, 92):

「孩子，你是從竹子里面生出來的，要記得，要回到我們的祖源地，跟自己相會，跟自己的家相會，跟自己的血源相會。」

Child, you were born of bamboos, you must remember, you must return
to our ancestral homeland, to meet yourself, to meet your own home, to
meet your own blood.

Perhaps the old man sensed what was about to happen. This theme
of "homecoming" is consolidated by a reappropriation of the "wise
Indigenous Elder" cliché, which, in this instance, serves to revitalise
prominent figures of Paiwan oral tradition. The old man embodies the
ancestors' spirit. When he encourages Dong-Qing to (re)discover his
Native roots, he initiates the hero's return process to his ethnocultural
identity—associated with bamboo groves, an integral part of Paiwan
landscape.

The Elder's wisdom and bamboos prove to be Dong-Qing and Xiao-
Ling's salvation. During the story's climax, the two are chased by White
Ring leader's ruthless enforcers for discovering the truth about the "incident". They manage to escape on a raft made of bamboos that suddenly
sprouted on the seemingly dead surface tilled by the old man, and that
became a tall grove within a few hours (*Ibid.*, 98–100). Once they finally
reach a faraway land, Suaiyung's ecological tale concludes with Xiao-
Ling's words: "Today, we, who were drifting, are about to land" (*Ibid.*,
100). Whether this is the homeland mentioned by the Elder or not is left
unsaid. Nevertheless, this couple is now free to "meet" its Indigenous
identity, liberated from the alienation feelings weighing them down, and
free to build a world of their own, founded on values contrary to White
Ring's non-organic societal model.

[Crimson Earth] reflects Paiwan People's concerns as they witness
the devastation of their territories and the gradual decrease of plant
and animal species, some, as with *Troides aeacus kaguya*, a butterfly,
and *Neofelis nebulosa brachyura*, the Formosan clouded leopard, to
the point of extinction. However, the story ends on a hopeful note
when Xiao-Ling and Dong-Qing escape—like the flood stories' surviving
couple—finding their way back to their Indigeneity. In this new space,
the People's memory can take on renewed forms, notably through the
literary productions of millennial authors Suaiyung and Ljivangerau.

LJIVANGERAU'S MULTICULTURAL TAPESTRY IN [MOVING. FORMOSA]

Tjinuay Ljivangerau, from the Kasuga tribe (*Chunri buluo* 春日部落), is another budding Paiwan author who employs the "returning home" motif—implying a return to Indigenous land, culture, language, and practices—in her poem [Moving. Formosa] (*Yidong. Fu'ermosha* 「移動·福爾摩沙」. It opens on a precolonial time when "Paiwan Elders said typhoons were a gift from Heaven" bringing bountiful fishing and other "rewards". Typhoons and "tropical cyclones" carried "fruits and seeds across islands", their "torrential rain" making them "fall, disperse and grow". Additionally, typhoons were channels of communication between man and nature, and between different nations, their "wind and rain" carrying "the greetings from Southern islanders" across isles (Ljivangerau 2010: 225).

Dichotomous Typhoons

Ljivangerau subsequently contrasts this natural equilibrium with "globalisation cyclones", and "far-reaching international markets typhoons" that arrived "at the end of the twentieth century". These unnatural storms "engulfed" migrant workers coming from "the Philippines, Vietnam, Thailand and Indonesia", who were used as "cheap labour" and forced, like the Paiwan, to leave their native villages for arduous and hazardous occupations (*Ibid.*, 225). Going back to her People, Ljivangerau recounts how, to avoid "being absorbed by inegalitarian trading markets", they retreat into happier memories, far from capitalist ideologies that transformed what was once a blessing into creophagous super-storms (*Ibid.*, 226–227):

在餘暉將盡黑暗漫襲部落的彼時	When the afterglow is about to disappear, and the tribe is immersed in darkness
我的族人也曾在最遠的遠洋魚船與最高不可攀的鷹架上	My tribesmen, who were also on the furthest deep-sea fishing boats and the highest unclimbable scaffoldings
從大腦皮質層的深處	From the far depths of their cerebral cortex
提領大武山盛開的百合花香與慶豐年的歌舞鈴鐺聲	Extract the perfume of lilies blooming on mount Dawu and the sounds of the Harvest Festival's songs, dances, and tinkling bells
[…]	[…]

我的族人掙扎於不平等交易市場的吸納
伺機脫逃

[...]
福爾摩沙是太平洋上季風與熱帶氣旋交織
密羅之處

吹散的種子與果實一落在這只有番薯大的
土地上

無不熱鬧了林相、豐富了物種 滋養這島
的豐腴

而移動的人群不能是種子嗎?也想用力呼
吸蹦芽發枝

我要趕在下一個暴風離潮分化我們之前

將那些因移動的剝離之傷

鑲嵌在你我交會的疊影上 成為我們美麗
的織錦

在生命幽微時閃熠 在無依軟弱時提醒

我們在福爾摩沙
都應 都要 扎根開花 處地花開

My tribesmen are struggling not to be
absorbed by inegalitarian trading markets
Waiting for the opportunity to escape

[...]
Formosa is a place where monsoons on
the Pacific Ocean and tropical cyclones
interweave and gather closely together

The dispersed seeds and fruits fall
everywhere on this sweet potato-sized
land

Invigorating the forests, enriching the
species
Nourishing this island's fertility

And can't moving crowds be seeds? They
also wish to breathe, spring sprouts, and
generate branches forcefully

I will hurry, before the next storm
separates us

To inlay these separation wounds caused
by moving

On the ghost superimposed images of
you and I To turn it into our beautiful
tapestry

To shine when life is obscure To be a
reminder in moments of isolation and
weakness

In Formosa
We all must We all will Take root and
blossom where we are

Like Monaneng, Awu and Sakinu decry dominant societies' treatment of
Taiwan Indigenous Peoples, Ljivangerau exposes the physical and psychological repercussions of the ongoing exploitation of her People by those
Awu introduced as "capitalists" in her story (1996: 39), and who, two
decades later, are still infringing on Indigenous Peoples' rights and lands.
Having found no other alternative to survive, Ljivangerau's tribesmen
contribute—albeit as "lowly" labourers—to the senseless urbanisation
depicted in Suaiyung's dystopian story. They are trapped into being actors
of the environmental destruction they fear. But Ljivangerau compares
them to "seeds" who "wish to breathe, spring sprouts and branch out
forcefully", calling upon their age-old resilience so they may persevere in
spite of current unnatural typhoons making them lose their footing.

"Weaving" An Epic Tapestry

Ljivangerau's writing—a pictural support of her People's (hi)story—is the medium she utilises to perpetuate, adapt, and reinvent the ancestral practices of storytelling and weaving. In fact, traditionally, the two were intertwined, since women weaving together would pass the time by telling each other stories (Awu 2016). In one of Awu's stories titled "Who Will Wear the Beautiful Clothes I Weave?", an Amis Elder is worried about the future of her People's customs, since most young people had left their native territories to be educated and to find work in larger cities (1996: 12–15). Likewise, the fate of Ljivangerau's tribesmen, who left home to work on boats and in cities, is a pivotal matter of the poem. Yet Ljivangerau, an urban writer who left home to study and then work in Taipei, reassures her people on the creative potential of movement, and even encourages them to reclaim the cities or any part of the territory on which they—like seeds carried by typhoons—land.

Ljivangerau thus ensures the permanence of Indigenous arts and follows in the footsteps of the precolonial Elders and early generations of Paiwan writers who preceded her. To guide the next generations, she is providing a literary map, or tapestry, "as a reminder in moments of isolation and weakness". A reminder of their past resilience in the face of climatic trials and colonisers' storming their land, their culture, and their very identity. If, in the words of Suaiyung, "man cannot live separated from the land" (2010: 89), or from their "tribe", stories of survivance have the potential to "heal" "separation wounds" by encouraging the island's Natives to "Take root and blossom" "In Formosa" (Ljivangerau 2010: 227).

Inspired by Typhoon Morakot, Suaiyung and Ljivangerau saw in this natural disaster the opportunity to affirm their tribal identity in literary works structured around a natural phenomenon. Although highly dangerous, it is not only an integral part of the People's life, but it also allowed them to develop their resilience. After being subjected to dehumanising colonial policies denying them their status of "civilised" men, the Paiwan turn to the nature that nurtured and protected them. In a federating perspective, Ljivangerau's poem and Suaiyung short story become Indigenous manifestos inciting all Peoples to fight colonial governments' unjust policies, be it assimilationist, human rights, or land and ecology related. These works consequently take their place amongst a committed Indigenous literature that serves as a marker of identity and carries on the cohesive function of oral stories, by reinforcing community ties in Taiwan and beyond.

"Ecological Tools" Inherited from the Ancestors

Suaiyung and Ljivangerau's narratives, in continuation of ancestral stories, are "renunciations of dominance, tragedy and victimry" (Vizenor 1999: vii), even when its heroes are facing the invasion of their mother-land. They detail the dire aftereffect of dominant societies' unbridled exploitation of the earth's resources and insist on the importance of a strong Indigenous identity grounded in values of cooperation between humans and all other species. Profiteering from Indigenous Peoples' state of vulnerability they themselves created, the colonisers encouraged the Paiwan to embrace capitalist values, making them complicit in their plunder of the island's natural resources. In response to this ongoing plight, established and budding authors extend their kinship ties interna-tionally to distant parents—oppressed minorities, uprooted casualties of dislocation, building bridges connecting Indigenous Peoples throughout the world. Together, they create platforms providing solutions based on their values of union and inclusion with nature in its entirety.

Building Bridges Between Communities

Paiwan authors have participated in international forums on Indigenous issues for decades. For example, in 1991, Awu attended the *International Indian Treaty Council* in Alaska, where Northern America's Indigenous Peoples engaged in discussions on ongoing struggles and personal expe-riences. In 2008, Sakinu's native village, a success in terms of Paiwan cultural revival, was visited by Indigenous Peoples from the Philippines and North America who had gathered in Taiwan to attend the *World Summit of Indigenous Cultures* (*Quanqiu yuanzhumin wenhua huiyi* 「全球原住民文化會議」). By being actively involved in such meetings, where conversations are centred around the breakdown of social and cultural systems, leading to a physical and mental estrangement from Native *locale* and traditions, the Paiwan set up means of protecting local and global ecology.

During the 1980s Indigenous Movements, intellectuals and activists who were defending their rights, including to self-determination, scru-tinised the 1960s and 1970s Red Power movement of North America

(Poiconu 2015). This influence from their Native American kin is present in the works of Paiwan authors where common experiences are emphasised. In his story, Suaiyung gives his own version of Crimson Earth's "Trail of Broken Treaties". Indeed, a year prior to the "incident", White Ring was at war with Crimson Earth and its scientists devised a project called "Gene Transcoding" (*zhuanma* 轉碼). White Ring leaders put it into effect even though they had signed a peace treaty with Crimson Earth, wreaking havoc on Crimson Earth's biome, and leading to the slaughter of two-third of its population (Suaiyung 2010: 95–96). And all those who set out to denounce these iniquitous acts were imprisoned or hunted down to be killed. It should be noted that Gu-Jiang, the village head who is divulging this "Gene Transcoding" project, characterises it as a "conspiracy" and "plot" (or "scheme",[8] *Ibid.*, 94–96), words used by those who believe climate change is a made-up conspiracy—but very much a reality for Indigenous People who suffer its effect, as with Typhoon Morakot.

Moreover, White Ring (*Baihuanquan* 白環圈, *huan* 環 meaning "ring; hoop", and *quan* 圈 "circle; ring") the apex of artificiality, of a milieu severed from its attachment to nature, is the antithesis of North American Indigenous Peoples' Sacred Hoop (or Sacred Circle), a Medicine Wheel divided in four colours (generally yellow, red, black, and white) embodying the four cardinal directions, elements, cycles of life, seasons, and core values. It symbolises tribal unity since everything—and everyone—in the universe is related.

By opposition, White Ring, white in part due to the fluorescents lights that illuminate its walls, is an aseptic universe harbouring a factitious "circle of life" where food is produced without soil, where seasonal and harvesting cycles no longer exist, and where night and day have become indistinguishable. In Suaiyung's story, the Sacred Hoop is broken, and the Elder is the only one trying to maintain a semblance of connection with the land and with ancient traditions. The heroes only survive by reconnecting with nature (the bamboos), with the ancestral stories told by the old man, and by joining forces. This is the message conveyed by Indigenous Peoples today, in literary and artistic works, in international political forums, in the media, or any stage where their voices can be heard.

---

[8] The author uses the terms *yinmou* 陰謀, "a conspiracy; plot; scheme"; and *pansuan* 盤算, "to plot; to scheme".

Ljivangerau's platform is her poem, in which she reminds her People that Formosa is a meeting ground for natural elements and for humans. The migrant workers who came to Taiwan are the same neighbouring islanders who used typhoons to send their greetings, related to the Paiwan "linguistically", and "members of the same tribe" (Ljivangerau 2010: 225). They form a large Indigenous family, moving "from island to island" and "from tribe to metropolis", "coming together, then spreading apart" sharing the same "fate". Ripped away from their homeland, they too try to escape their current reality of "discrimination" and unfair governmental "policies" by taking their "homesick souls" back to their native "fragrant paddy fields" (*Ibid.*, 226). Now in Taiwan, along with its lush forests and species diversity, they are part of a colourful cultural tapestry, and are the island's greatest wealth.

Ljivangerau herself, through her "tapestry" metaphor and her appeal to all Indigenous Peoples to "blossom", joins a global Indigenous family of women storytellers, weaving "together in a fabric of interconnection" (Gunn Allen [1986] 1992: 11). She continues along the path laid out by "elder sisters", like Laguna Pueblo/Lakota and Lebanese-American scholar and author Paula Gunn Allen (1939–2008) who, in 1992, lauded Indigenous Peoples' resilience and ability to blossom ([1986] 1992: xi):

> But it is of utmost importance to our continuing recovery that we recognise our astonishing survival against all odds; that we congratulate ourselves and are congratulated by our fellow Americans for our amazing ability to endure, recover, restore our ancient values and life ways, and then blossom.

"Pay Attention to Indigenous Customs"

In both Taiwan and the United States, colonisation is still a reality for Indigenous Peoples, the "determined guardians of [their] motherland, as [they] have been for thousands of years, and will continue to be" (Poiconu et al. 2019).[9] Fighting for their right to self-determination,

---

[9] This is an excerpt from the January 2019 open letter by Indigenous representatives serving on the Taiwanese Indigenous Historical Justice and Transitional Justice Committee, in response to China's president Xi Jinping claim on the 2nd of January 2019 that Taiwan was "historically" and "legally" part of China. It was translated in English by the *Georgia Straight*, and is available in the original Mandarin on the *Apple Daily* website (published on the 8th of January 2019) at: https://tw.appledaily.com/forum/20190108/D2XNZZXLFYT7V7E46O2GYRLOBQ/.

they never gave up claiming their territory in its entirety. Ljivangerau uses the name Formosa (given to the island not by former colonial rulers but by Portuguese explorers), instead of Taiwan or the "Republic of China", Han designations by no means axiomatic for the isle's Indigenous Peoples. It shows the Paiwan never renounced their tribal and cultural sovereignty, despite various policies and projects by the Republic of China aimed at the edification of a Taiwanese national identity.

As wars of "imperial conquests" are not "solely or even mostly waged over the land and its resources", but are "been fought within the bodies, minds, and hearts of the people of the earth for dominion over them" (Gunn Allen [1986] 1992: 214), Ljivangerau is dissociating from Formosa's prominent or official Han appellations. She is defiantly resisting imperialist encroachment within the mind of the Indigenous Peoples—what Awu regards as an "ideological colonisation", or "colonisation of the mind"(*sixiang shang de zhimin* 思想上的殖民, Awu [1997] 2001: 45).

In [Crimson Earth], Suaiyung provides clues concerning the origins of these wars of imperial conquests—the oblivion of our common humanity. The old man recounts how, "a long time ago", two brothers fled their overpopulated homeland. They stumbled upon the Crimson Earth, which was so small they decided the eldest would live underground, and the youngest on the land, neither depriving the other of his resources. However, their descendants "forgot this blood relationship" and started to fight, so they were "punished" by the "gods" (2010: 98). Suaiyung links the two opposed societies by kinship and a history of exile, reminding us of the challenges humanity is now facing, encouraging us to listen to Indigenous Peoples' consideration of climate change.

In this matter, Paiwan People again turned to their international Indigenous family. Together they generated the *International Indigenous Peoples Forum on Climate Change* in 2008, "as a caucus for indigenous peoples participating in the UNFCCC [United Nations Framework Convention on Climate Change] processes" (IIPFCC 2008). Their website puts forward Indigenous perspectives and knowledge of sustainable practices in their use of natural resources—acquired after a long and conscientious observation of their milieu and based on an almost symbiotic relationship with nature (IIPFCC 2008):

Indigenous peoples (IPs) have a particular contribution to make in discussions around climate change and sustainability, given their strong historic and cultural connection and the stewardship role they continue to play in sustainably managing many of the world's biological resources.

With events like Typhoon Morakot, the world's political leaders and decision makers can no longer ignore the pressure from human activity on the biosphere and could heed Indigenous Peoples' advice. For instance, Ojibwe writer and green activist Winona LaDuke (1959–), concluded her book on North American Indigenous communities' environmental fights, *Recovering the Sacred: The Power of Naming and Claiming*, by drawing attention to Native Peoples propensity to offer solutions to global climate change (2005: 253):

> Native American communities are creating momentum for change and providing some critical leadership in the face of global climate change and the energy crisis to come. By democratising power production, Native nations are providing the solutions that all of us will need in order to survive into the next millennium.

Twenty years ago, nine years prior to Morakot, Tsou (*Zou zu* 鄒族) Professor Pasuya Poiconu (Pu Chung-cheng 浦忠成) wrote an article published in the *Taipei Times* titled "Pay attention to Aboriginal customs". It was penned in response to the Taiwanese government's allegations that Indigenous hunters were "destroyers" "inimical to wildlife preservation". Poiconu's rebuttal addressed the fact that if some Indigenous individuals were hunting irresponsibly, it was due to centuries of colonial policies—especially those forbidding them to enter their traditional "hunting, fishing and food-gathering grounds"—forcing them to "ignore tribal taboos and ethics" "to survive". According to Poiconu, traditional hunting was, on the contrary, a "manifestation of Aboriginal beliefs and ways of life, including religion, preservation of the land, distribution of resources, tribal discipline and taboos". He then pointed out that "centuries of Aboriginal hunting and land reclamation did not drive Taiwan's wildlife into extinction, nor have any other serious consequences such as extensive mudslides", holding the government accountable and imparting ideas applicable to countries throughout the world (Poiconu 2000):

> The most effective way to love wildlife and protect the ecology is to stop road construction across green areas, stop building unnecessary dams, ban mining activities, reduce betel nut farming, punish the vendors of hunted game, build fish channels across check dams and similar measures.

The future of Indigenous literature could partly reside in its impact on our "hyperconnected" societies who developed complex industrial and digital "ecosystems" yet failed to foster our link to vital ecosystems. Suaiyung and Ljivangerau are handing us a mirror reflecting our condition as "modern" humans, conceivably condemned to be crushed by the globalisation juggernaut. Non-Indigenous people who did not grow in resilience could benefit from paying attention to Indigenous customs and learn to "take root and blossom" "where [they] are" (Ljivangerau 2010: 227).

Conclusion

> The past space and lifestyle were both destroyed. The land was exploited and broken. If we could not trace back to the historical contexts, pursue the ancestors' footsteps, learn from their experiences and wisdom, and establish the substantial relationship and identification based on the deepest land tradition and ethics, though we may have completed tribal maps, could we still boast about returning to the land of the ancestors? Actually, myths do not lie. The land has inlaid brilliant jewels for us. They were the firm support of our assertion.
>
> Pasuya Poiconu (2012: 304–305)

With international organisations such as the UNESCO now insisting on Indigenous Peoples' rights, environmental and traditional knowledge policies, academia is opening more and more to Indigenous-oriented studies. Yet the scope of UNESCO's policies is limited, as they often fall short of their goals. Scholars, for their part, have an ethical responsibility towards studied Indigenous communities, who advocate for academic recognition of their own epistemologies, to shift the discourse from "objects" of study to active "subjects" of their own narratives. They must accept these "twin requirements": "to enter tribal philosophies and to enter tribal relations", so as to "build bridges between Native communities and the academy" (Garroutte 2003: 110).

All of humanity could benefit from Indigenous knowledge, first by accepting its validity, then by applying it. Some universities in Taiwan, like Tsing Hua University 國立清華大學, with professor Fu Liyu 傅麗玉—who has not only worked with Indigenous tribes in Taiwan for over twenty years, but has also travelled to learn from tribes in Alaska and South America—are promoting Indigenous life sciences and technology, including environmental conservation methods, and have been developing science education programmes for over two decades. In fact, the University's president, Hochen Hong 賀陳弘, admitted that Indigenous Peoples' "important scientific knowledge has been largely overlooked" (National Tsing Hua University 2021). That is why, in 2019, the University created Taiwan's first Center for Indigenous Science Development with Fu as its director, and, in conjunction with the Indigenous Culture and Education Programme of the Taiwanese Ministry of Education, started construction in 2021 on a pavilion to house said Center as of 2022.

According to Hochen, the Center was created "in order to deepen our understanding of the traditional wisdom of Taiwan's indigenous peoples, and to find ways in which it can be applied to finding effective solutions for such pressing issues as global warming and climate change" (National Tsing Hua University 2021). Such initiatives could be replicated in universities all over the globe. But first, we must accept the reality of climate change, whilst recognising and utilising Indigenous expertise. A novel way to approach perceptions of climate change can be through Indigenous literary works that convey a sense of urgency when dealing with this issue. This literature is also imbued with community and solidarity ethics presented as imperatives of both long-term human and planetary survival.

As I am concluding this paper, the *Intergovernmental Panel on Climate Change* just published an alarming report. Written by two-hundred and thirty-four experts from sixty-six different countries, it states that scientists are observing unprecedented human-induced climate changes across the globe and estimate that we only have a short window of time to modify our behaviour (IPCC 2021). This confirms what Paiwan authors have been admonishing us about. Through personal and fictional stories, reminiscent of oral tradition, they have maintained their inclusive and eco-friendly values alive. By sharing them with the world, they not only alert us about the issues that impede their rights and their liberties, but also provide us with examples of ecological sustainability and social justice, according to their People's relational model of harmony with nature and all its denizens.

REFERENCES

Awu, L. 阿(女烏)·利格拉樂 (1996). *Shei lai chuan wo zhi de meili yishang* 誰來穿我織的美麗衣裳 [Who Will Wear the Beautiful Clothes I Weave], Taizhong 台中: Morning Star 晨星.

Awu, L. 阿(女烏)·利格拉樂 (1997). *Hong zuiba de VuVu* 紅嘴巴的VuVu [Red-mouthed Vuvu], Taizhong 台中: Morning Star 晨星, 2001.

Awu, L. 阿(女烏)·利格拉樂 (2016). Unpublished interview conducted by Fanny Caron-Scarulli, 27 June.

Bourguignon, L. and Bourguignon, C. (2015). 'La mort des sols agricoles', *Études sur la Mort*, 148 (2), pp. 47-53, https://doi.org/10.3917/eslm.148.0047.

Chang, JC. and Slaymaker, O. (2002). 'Frequency and Spatial Distribution of Landslides in a Mountainous Drainage Basin: Western Foothills, Taiwan', *CATENA*, 46 (4), pp. 285-307, https://doi.org/10.1016/S0341-8162(01)00157-6.

Chen, CY. (2016). 'Landslide and Debris Flow Initiated Characteristics after Typhoon Morakot in Taiwan', *Landslides*, 13 (1), pp. 153–164, https://doi.org/10.1007/s10346-015-0654-6.

Chiang, B. 蔣斌 (ed.) (1920). *Fanzu guanxi diaocha baogaoshu diwujuan, paiwanzu diyice* 番族慣習調查報告書第五卷. 排灣族第一冊 [Investigation Report on Barbarian Tribes' Customs, Volume 5: The Paiwan, Book 1], Taipei 臺北: [Academia Sinica's Ethnology Institute] 中央研究院民族學研究所, 2003.

Early, R. and Whitehorn, J. (eds.) (2003). *One hundred Paiwan texts*, Canberra: Research School of Pacific and Asian Studies; The Australian National University.

Garroutte, EM. (2003). *Real Indians: Identity and the Survival of Native America*, Berkley; Los Angeles; London: University of California Press.

Gunn Allen, P. (1986). *The Scared Hoop: Recovering the Feminine in American Indian Traditions: With a New Preface*, Boston: Beacon Press, 1992.

IIPFCC (2008). *About the International Indigenous Peoples' Forum on Climate Change*. Available at: https://iipfcc.squarespace.com/who-are-we-1 (Accessed: 18 June 2021)

IPCC (2021). 'Climate Change Widespread, Rapid, and Intensifying—IPCC', *IPCC: News, Press Release* [Online]. Available at: https://www.ipcc.ch/2021/08/09/ar6-wg1-20210809-pr/ (Accessed: 9 August 2021)

Jet Propulsion Laboratory/NASA (2008/Last updated July 20 2021). 'Global Climate Change: Causes', *NASA Global Climate Change and Global Warming: Vital Signs of the Planet* [Online]. Available at: https://climate.nasa.gov/causes/ (Accessed: 21 February 2021)

Jeunehomme, A. (2018). 'Lydia et Emmanuel Bourguignon', *Openfield*, 11 [Online]. Available at: https://www.revue-openfield.net/2018/07/03/lydia-et-emmanuel-bourguignon/ (Accessed: 21 March 2021)

Laduke, W. (2005). *Recovering the Sacred: The Power of Naming and Claiming*, Cambridge: South End Press.

Ljivangerau, T. (Chen, MJ. 陳孟君) (2010). 'Yidong. Fuermosha 移動. 福爾摩沙' [Moving. Formosa] *in* Sun, TC. 孫大川 (ed.) *Yong wenzi niangjiu. Yong bi lai changge: 99 nian taiwan yuanzhuminzu wenxue jiang de jiang zuopinji* 用文字釀酒. 用筆來唱歌: 99年臺灣原住民族文學獎得獎作品集 [Using the Written Language to Make Wine, Using a Pen to Sing: 2010 Taiwan Indigenous Literary Awards Winning Works' Collection], Taipei 臺北: Council of Indigenous Peoples 行政院原民會, pp. 224-227.

Monaneng 莫那能 (1989). *Meili de daosui* 美麗的稻穗 [Beautiful Ears of Rice], 2nd Reprint, Taizhong 台中: Morning Star 晨星, 2014.

National Tsing Hua University (2021). 'A New Base for the Center for Indigenous Science Development', *National Tsing Hua University: News*, 07 September [Online]. Available at: https://nthu-en.site.nthu.edu.tw/p/406-1003-214403,r4806.php (Accessed 03 December 2021)

Poiconu, P. (Pu, C. 浦忠成), Huang, F. (Translation) (2000). 'Pay Attention to Aboriginal Customs', *Taipei Times* (Editorials), 13 December [Online]. Available at: http://www.taipeitimes.com/News/editorials/archives/2000/12/13/0000065331 (Accessed 2 November 2015)

Poiconu, P. (Pu, C. 浦忠成), Wordsworth (Translation) (2012). *Literary History of Taiwanese Indigenous Peoples (Volume 1)*, Taipei: National Academy for Educational Research; Le Jin Books Ltd. 里仁書局.

Poiconu, P. (Pu, C. 浦忠成) (2015). 'Suizhe shiguan bianhua er zhuanzhe de yuanzhumin wenxue 隨著史觀變化而轉折的原住民文學' [Indigenous literature in the wake of historical outlook changes and turning points], *Yuanzhuminzu wenxian* 原住民族文獻, 24 [Online]. Available at: https://ihc.cip.gov.tw/EJournal/EJournalCat/283 (Accessed 6 February 2021)

Poiconu, P. (Pu, C. 浦忠成), Mateli Sawawan, (馬千里), Magaitan Lhkatafatu *et al.* (2019). 'Indigenous peoples of Taiwan tell Xi Jinping that they've never given up their rightful claim to sovereignty' [English translation of a joint statement], *Georgia Straight*, 17 January [Online]. Available at: https://www.straight.com/news/1189736/indigenous-peoples-taiwan-tell-xi-jinping-theyve-never-given-their-rightful-claim (Accessed 15 November 2020)

Sakinu, A. 撒可努·亞榮隆 (1998). *Shanzhu. Feishu. Sakenu (xiuding ban)* 山豬. 飛鼠. 撒可努 (修訂版) [Boar. Flying Squirrel. Sakinu (Revised Edition)], Taipei 台北: Yelu耶魯, 2011.

Sakinu, A. 撒可努·亞榮隆 (2002). *Shanzhu. Feishu. Sakenu 2 Zoufeng de ren* 山豬. 飛鼠. 撒可努2走風的人 [Boar. Flying Squirrel. Sakinu 2: Wind Walkers], Taipei 台北: Yelu耶魯, 2011.

Simon, S. (2002). 'The Underside of a Miracle: Industrialization, Land, and Taiwan's Indigenous Peoples', *Cultural Survival Quarterly*, 26 (2) [Online]. Available at: https://www.culturalsurvival.org/publications/cultural-survival-quarterly/underside-miracle-industrialization-land-and-taiwans (Accessed: 16 May 2021)

Streiff, L. (2021). 'NASA Satellites Help Quantify Forests' Impacts on Global Carbon Budget', *NASA Global Climate Change and Global Warming: Vital Signs of the Planet* [Online]. Available at: https://climate.nasa.gov/news/3063/nasa-satellites-help-quantify-forests-impacts-on-global-carbon-budget/ (Accessed 21 February 2021)

Suaiyung, I. 索伊勇·以新 (Zhu, KY. 朱克遠) (2010). 'Chitu 赤土' [Crimson Earth] *in* Sun, TC. 孫大川 (ed.) *Yong wenzi niangjiu. Yong bi lai changge : 99 nian taiwan yuanzhuminzu wenxue jiang de jiang zuopinji* 用文字釀酒. 用筆來唱歌: 99年臺灣原住民族文學獎得獎作品集 [Using the Written Language to Make Wine, Using a Pen to Sing: 2010 Taiwan Indigenous Literary Awards Winning Works' Collection], Taipei 臺北: Council of Indigenous Peoples 行政院原民會, pp. 224–227.

Sun, TC. 孫大川 (2009). 'Taiwan History and the Indigenous People 台灣史與原住民', *Taiwan Literature: English Translation Series* 台灣文學英譯叢刊, 24, pp. 9–13.

Taiban, S. 台邦·撒沙勒 (1993). 'Feixu guxiang de chongsheng: Cong "gaoshan qing" dao buluozhuyi: Yige yuanzhumin yundong zhe de guancha he fanxing 廢墟故鄉的重生: 從《高山青》到部落主義: 一個原住民運動者的觀察和反省' [Rebirth of the Native Place in Ruins: from *Mountain Greenery* to Tribalism: Observations and Reflections on an Indigenous Movement], *Taiwan Historical Materials Studies* 台灣史料研究, 2, pp. 28–40.

Taiban, S. 台邦·撒沙勒 (2006). The Lost Lily: State, Sociocultural Change and the Decline of Hunting Culture in Kaochapogan, Taiwan. Doctoral Thesis. University of Washington.

Tung, CF. 童春發 (1995). The Loss and Recovery of Cultural Identity: A Study of the Cultural Continuity of the Paiwan, a Minority Ethnic Group in Taiwan 文化アイデンティティの喪失と再生: 台湾原住民パイワン族の文化継続の研究. Doctoral Thesis. International Christian University 国際基督教大学.

Vizenor, G. (1999). *Manifest Manners: Narratives on Postindian Survivance*, Lincoln: University of Nebraska Press.

Wang, CC., Tseng, LS., Huang, CC., *et al.* (2019). 'How much of Typhoon Morakot's Extreme Rainfall is Attributable to Anthropogenic Climate Change?', *International Journal of Climatology*, 39 (8), pp. 3454-3464, DOI: https://doi.org/10.1002/joc.6030.

Younging, G. (2018). *Elements of Indigenous Style: A Guide for Writing By and About Indigenous Peoples*, Edmonton: Brush Education.

Climate Change, Humility, and Resilience: Analysing a Myth of the Bunun in Taiwan

Dean Karalekas and Tobie Openshaw

There have been many studies of indigenous land-use practices and calls for their incorporation into governmental green policies to aid in the effort to deal with the global threat of climate change. Much is being written on the subject, examining the practices of indigenous groups from every corner of the world, and how their Traditional Ecological Knowledge (TEK) can inform and improve upon existing environmental policies. For example, researchers such as Quintana-Ascencio et al. (1996), Thomas (2003), and Niamir-Fuller (1998) have studied such diverse groups as Mayan farmers, Hewa hunters, and herders in the Sahel region of West Africa to better understand disturbance ecology for the betterment of modern landscape design and game management practices. Moreover, Costa-Pierce (1987), Pearce (1993), and Berkes (1985) have examined such TEK systems as Hawaiian aquaculture, Caribbean forestry, and game management among Arctic peoples in order to learn lessons for avoiding the so-called tragedy of the commons (Martin et al. 2010).

D. Karalekas (✉) · T. Openshaw
Centre of Austronesian Studies, University of Central Lancashire, Taipei, Taiwan
e-mail: dkaralekas@uclan.ac.uk

© The Author(s), under exclusive license to Springer Nature Switzerland AG 2022
N. J. P. Alsford (ed.), *Pacific Voices and Climate Change*,
https://doi.org/10.1007/978-3-030-98460-1_7

147

Less attention, however, is being paid to how these practices were derived. This chapter therefore contributes to the literature on the epistemology of TEK which, having evolved over thousands of years, encompasses worldviews very different from the Western-dominated, rationalist perspective. We argue that a direct line can be drawn from indigenous belief systems to their experiential relationships with the natural world as retold via the narrative of their myths. The purpose of this chapter therefore is to examine that mechanism by looking at one particular myth from the belief system of one particular indigenous ethnic group—the Bunun nation of Taiwan—and to take a deep look at how this myth contributes to the formation of a worldview that places the human experience in balance with the ecosystem of which the Bunun themselves are a harmonious part.

Traditional Ecological Knowledge refers to practices and beliefs pertaining to the relationship that human beings have with the natural world, and with each other, and how these beliefs have evolved over thousands of years through intimate contact with the environment. This chapter builds upon definition of TEK employed by Kyle Powys Whyte (2013), who argues that the concept is a collaborative one in which various peoples and cultures borrow from one another as they conceptualize how they arrive at knowledge, and how this collaboration and on-going synthesis in modes of understanding has the potential to provide the tools needed to address the issue of climate change and serve as better stewards of the planet's natural resources. Sometimes referred to as Native Science or Indigenous Knowledge, TEK incorporates an understanding of the interactions between flora and fauna, between animals, humans, and landscapes, and between all of these and the seasonal progression of time, and how this all contributes to sustainable patterns of living within a specific habitat (e.g. with respect to more efficient techniques of hunting, trapping, fishing, and farming).[1] It is through the use of rituals and ceremonies that TEK provides its insights; encompassing the idea of reciprocity and of gratitude, as it puts a great emphasis on spirituality and respect for the sacred—concepts that are passed down through the medium of a culture's mythology (Rinkevich et al. 2011; Berkes 2012).

---

[1] For a link to the wider studies on use the TEK in climate and ecological change, see *Sacred Ecology* by Fikret Berkes (2012). In it, the author provides several case studies of sustainable land-use practices by indigenous cultures in regions as varied as tropical rainforests and the arctic.

This chapter begins with a look at the importance of mythology as a form of narrative, and the central role it plays in crafting and perpetuating a TEK worldview. It then provides a retelling of the Bunun myth of the Shooting of the Sun, as published in various ethnographies and as relayed to the authors in personal interviews with indigenous knowledge-holders not just on the myth itself, but on its importance to the culture and identity of the teller (Miciq 2021; Yumin 2016). Next, we take a brief look at the universality of this myth, and how it is a common theme across cultures. We then examine a few theories that have been offered to explain how this myth may have emerged. From here, we look at the Bunun conception of myth telling, or "speaking of the past," and the importance this played in their day-to-day lives in the pre-Christian era, followed by a brief examination of the Christian influence on the Bunun communities and their worldview. Next, this myth is analysed using a psychological and sociological approach to determine the cultural significance of the Shooting of the Sun myth, and how it plays a role similar to myths of other cultures around the world; including how it represents a paradigmatic synthesis marking the shift from a hunter-gatherer economy to an agrarian economy, made possible by the progressive tension of the sun/moon dyad. Finally, we look at what larger lessons the Western world may learn from the Bunun's example.

The Importance of Myth

Myths are far more than mere entertainment: they are the channel through which the elders of a cultural group instil values and inculcate members of the younger generation into that society's belief system. They are an essential medium for the transmission of ways of knowing, particularly with respect to TEK, and conveying their perspective on man's place in the natural world.

In preliterate societies, the practice of oral storytelling has served to preserve important myths and hence the truths they contain. They are essential in building and passing on knowledge about the internal and external world, especially the natural world. Indeed, the godfather of comparative religion, Max Müller, believed all myths were nature myths, offering an explanation for the natural world in the manner of a proto-science (Müller 1856; Segal 2016; Kirk 1984). The importance of myths to the survival of culture thus cannot be overstated. If, per McNeill (2013), culture can be defined as the body of knowledge and beliefs that must be held by its members, then myths are the manner in which these beliefs are transmitted to new members and new generations. Hence, the veracity of this narrative is immaterial: according to Benedict Anderson (2006), even a fictitious mythology can contribute to building an "imagined community" that helps a group become a community. In Anderson's words, the nation is "imagined as a community, because, regardless of the actual inequality and exploitation that may prevail in each, the nation is always conceived as a deep, horizontal comradeship" (McNeill 2013; Anderson 2006).

Mythologist Joseph Campbell (2004) identified the four functions that myths serve in a culture: The first is to invoke gratitude and awe in the mystery of existence. The second is to provide a heuristic for understanding the world around us and our place therein. The third is to provide a framework for a system of morality and propriety that allows society to function smoothly. The fourth function of myths is to serve as a psychological roadmap to navigate through the various stages of life.

The second and third functions of myth have, since the age of Enlightenment, been taken over by rationality and the scientific method. In the Western knowledge traditions, no longer do we find answers to our understanding of the world (and hence our place in it) from religious belief systems, but from science and, increasingly, politics. During the Enlightenment, the mystical components of systems of knowledge acquisition were stripped away in favour of a dogmatic focus on the rational, forcing knowledge systems that touched upon the spiritual to either go underground, living on only in secret societies and mystery cults, or be transferred to the exclusive purview of organized religion. So it is that Alchemy transmuted into Chemistry, and Astrology into Astronomy. Despite the undeniable benefits of living in a civilization whose motivating epistemology is derived from the Enlightenment, man does not live by logic alone, and the spiritual aspect of being that has been removed from the accepted mainstream systems of scientific and rational knowledge has left a gap in the human experience (Pinker 2018; Gare 2001; McIntosh 2012).

Of course, there is still traditional knowledge, superstitions, and traditions defining the way agriculture and everyday life should operate in Western societies, especially at the community level—just as surely as not all indigenous people live in harmony with the natural environment. Those who do not, however, can in many cases be said to be living a "Western" lifestyle. Hence, this chapter focuses on the paradigms; not the peoples who hold them. But just as global problems demand globally coordinated actions, a scientific approach to solving them is the easiest to rationalize across nations and across cultures, and this has risen to the fore. As a result, in the West, climate change is framed as a problem created by humans that must be solved by humans, through the medium of our technology and shifting our consumptive lifestyle, because as humans we have agency and an outsized impact on the planet. Evidence of the near-exclusivity of this paradigm can be found throughout the climate-change literature distributed by international nongovernmental organizations (NGOs) and transnational governance bodies such as the European Union and United Nations, and particularly the ultimate authority on such matters, the Intergovernmental Panel on Climate Change (IPCC), which according to its website (https://www. ipcc.ch/), was created specifically to provide policymakers with regular scientific assessments on climate change, which, according to the IPCC Working Group I report, *Climate Change 2021: The Physical Science Basis*, approved by the 195 member governments of the IPCC, is unequivocally caused by human activities, and is unequivocally affecting every corner of the Earth (Zhai et al. 2021).

As we move forward beyond the postmodern era, however, we are discovering that our reliance on such secular institutions and explanations is unfulfilling. For one thing, it does not address the spiritual needs that every human being experiences as a consequence of consciousness. For another, science is, by design, lacking in certitude: The entire purpose of the scientific method is to challenge the existing state of knowledge for the dual purpose of testing its integrity, and honing, improving, and building upon it. It is therefore constantly in flux. Politics is even less reliable as a source of social stability, and in fact often works contrary to that end.

By contrast, mythological systems—and the traditional cultures built upon them—evolved over thousands of years to serve the ends of cosmogony and of ordering society, and we dismiss them at our peril. Critics who read mythology as literal and dismiss it for its impossible events ignore the deeper allegorical and metaphysical truths contained therein, and do both the individual and society a disservice by sidelining these ancient narratives in favour of the fleeting beliefs of the day. Scientists are slowly coming to the conclusion that traditional societies with a strong sense of their symbiotic interconnectedness with the natural world have much to teach us about how to address many of the problems we face today, including resource depletion and a changing climate. Hung (2013) therefore proposes a fifth function of mythology: namely, an ecological function, which speaks to the relationship that humans enjoy with the interconnected natural world they find themselves in, and encompassing other living beings, plants, water courses, the ever-changing seasons, and the spirit realm. The Bunun tale of the Shooting of the Sun is an excellent example of myth fulfilling this function.

Shooting of the Sun

In the time of *the Great Chaos*, the world was not as it is today. There were two suns in the sky, and as one set, the other would rise. They took turns shining constantly upon the earth, meaning there was no night, and no relief from the heat. Because of the great heat produced by the two suns, millet could not grow well, and a drought beleaguered the land. Many animals died. Many people died.

One day, a family went to tend to their crops, as they did every day, and the mother placed their baby in the field, making a shade structure to protect the infant. After working for some time, the parents went back to check on their baby, only to find to their horror that the heat of the suns had killed the child and turned it into a lizard.

Terrified and enraged, the man vowed revenge: since the sun killed our baby, then I will kill the sun! Packing for a long journey, the man and his eldest son prepared for their mission of vengeance. Before setting out, they went to mark the spot where the baby had died by planting a tangerine tree.

The journey took many years and covered great distances. The son grew into a strong, clever young man, and his father taught him the skill of the bow. Finally, they came to a place in a strange, unknown land. It was a high plateau with a good view, and very close to the sun. Here, they would set up their ambush.

At last, their wait was over, and when the suns were low on the horizon, and the man could line up a good shot, he raised his bow and drew it back, waiting for the exact right time to strike. He loosed his arrow, and it arced through the sky, but it was no good: the arrow missed its mark. Quickly nocking another, the man prepared to shoot again, only the sun had risen higher in the sky, and the great heat and light now prevented him from aiming. He wiped the sweat from his brow, worried about missing his last chance to gain revenge for the death of his baby.

Seeing his father's distress, the son leaped into action. He unsheathed his machete and cut a frond from a nearby tree, which he used to shade his father from the sun's heat and light. This gave the man just the opening he needed to pull back on his bow and loose his second arrow. This time, his aim was true, and it struck its target, the eye of the second sun.

Blood gushed out from the sun's injured eye, and the blood became stars. The sun, now in great pain, asked the man for something with which he could wipe clean its pierced eye. The man threw aloft his cloth, whereupon the injured sun lost its fire, and turned cold and gray until it adopted the soft glow of what we now know as the moon.

The injured sun, which had now become the moon, was angry at the man and retaliated. It grasped him and demanded to know why he had done such a thing. "Because you killed my baby!" the man retorted, to which the moon replied, "Foolish man! Do you not know that your baby's death was your own fault? It was your incestuousness and ingratitude that caused him to die! It is only through my beneficence that you have life. Everything grows out of my favour. Yet, for all this, your people have never thanked me. You have not once thought about holding a ritual in my honour."

The man saw the error of his ways, and the moon released him. The man promised to follow the dictates handed down by the moon.

"This shall be our agreement," said the moon. "From now on you will have to toil very hard to earn your living. You will begin work when the cock crows early in the morning, and you cannot return home until you see my face. You must observe my comings and goings, and in return I shall protect your harvests."

The man promised to honour this agreement, and the moon left to heal its wounds. Thus the moon gave the man *Samu*—the rituals and taboos of the Bunun people, to be passed down from generation to generation.

The man and his son headed back for their long journey home. Although it took many years, the voyage back was far less arduous. Now, with only one sun in the sky, they could travel by day and rest at night. Trees and flowers grew more lush in the absence of the oppressive heat, and by the time they returned home, the tangerine tree they had planted was heavy with fruit.

As he had promised, the man delivered the moon's decrees to his people. He told them if they obey these rituals, their crops and their offspring would be abundant and strong. He warned them that if they did not, their tribe would surely die (Miciq 2021; Du and Winkler 2003; Karalekas 2013; Ho 1971; Fang 2016; Kuhlmann 2019).

A COMMON THEME

There are variations of this myth among the five distinct communities of Bunun. For example, in some tellings, the baby is replaced by a centipede, and in some, the man goes on his journey alone. Moreover, this myth is not restricted to the Bunun; The tale of a culture hero whose tribe is beset by the oppressive heat of a primordial world with two suns, and who shoots one of the suns to bring about the solar/lunar pattern we know today, can be found among the cosmological myths of no fewer than six of the Formosan aboriginal groups. There is a remarkable consistency in the details of these tales, albeit with some deviation. For example, in the Truku telling, it takes three generations to complete the quest: the man sets out on his journey with his son, and his grandson completes the mission many years later. Only that of the Puyuma deviates significantly, wherein there are initially eight suns, all but two of which (those that become the lone sun and the moon) are shot down by the culture hero (Kuhlmann 2019; Ho 1971; Miciq 2021).

Furthermore, the general theme of this myth is common outside of Taiwan as well, being shared among several cultures throughout the world. The Aztec creation myths speak of five suns—stand-ins for earlier Mesoamerican cultures—that were symbolically destroyed by deity figures, leaving the Aztec with the responsibility of feeding the current sun with *tlaxcaltiliztli*—nourishment derived from offerings of blood and hearts. In parts of China, the hero archer who shoots down nine of ten suns in the sky is named Hou Yi. The Japanese tell the tale of a mole who shoots down six of seven suns. The Miao people of the Pearl River basin have the story of Hsangb Sax, who climbs up a tree to shoot down superfluous suns. The Aruá people, an indigenous group in Brazil, tell of their own Great Chaos, when the heat was so intense that it burned a child, prompting the Aruá men to kill many suns in revenge. Their warring continued until a demiurge named Paricot lifted the heavens far above the Earth and out of man's reach. Other examples can be found in Borneo, Sumatra, and India (Metevelis 2005; Sweeney 2013; Mindlin 2002).

The author first heard a version of this myth recounted by Mongolian herders in the Gobi Desert. In that tale, the primordial world had seven oppressive suns in the sky. The community appealed to its greatest archer for help. Erkhii Mergen accepted the mission, arrogantly vowing to cut off his thumbs should he fail. One by one, he dispatched the suns, shooting from galloping horseback. His last arrow, shot at the seventh and final sun, was intercepted by a swallow, and so missed its mark, cleaving the bird's tail in two. True to his word, Erkhii Mergen cut off his thumbs and became the marmot, living in a hole in the ground and eating only dry grass. Each morning and evening, the marmot emerges from his burrow and ponders the sun, forgetting for a moment he is no longer the great Mongolian archer who shot down the six suns.

POSSIBLE ANCIENT OCCURRENCE

Like the widespread story of the Great Flood—which is shared by so many disparate cultures that some historians and archaeologists suggest it may refer to actual events or climatic processes, such as a post-glacial sea-level rise or a giant tsunami triggered by a meteor impact—the ubiquity of the superfluous suns as a mythological theme may hint at a similar ancient occurrence.

A NASA scientist, Chinese professor Zhao Feng, offered a hypothesis for just such a prehistoric event involving a comet striking the Earth. Such a comet would have broken apart due to tidal forces as it approached our gravity well, and the pieces would have rained down over a period of several days, burning up in our atmosphere in frightening, sun-like balls of flame. The collision of comet Shoemaker-Levy 9 with Jupiter in 1994 serves as an example of how this phenomenon could have unfolded. Professor Zhao's suggestion is consistent with Napier and Clube's (1979) planetary impact hypothesis—sometimes called the Shiva hypothesis—adding the observation that such a cataclysmic cosmic event could have been interpreted by our distant ancestors as the emergence of many suns in the sky, perhaps prompting panicked attempts to shoot them down—attempts that would have appeared successful once the comet fragments finally fell to Earth, one after the other. Although there is no evidence to support this playful hypothesis, it is an intriguing postulation to ponder (DayDay News 2019).

Another interpretation of the genesis of this myth is offered by American paleographer Sarah Allen, who suggests the story is a disguised history. In this ancient euhemerism, each of the suns is a totemic representation of one of the ten groups of Shang Dynasty (1600–1046 BC) elites. (Allan 2010; Kirk 1984). The ancient philosopher *Mo Tzu*, in an eponymous text dating to the fourthcentury BC, details a cosmic and climatic upheaval that predated the founding of the Shang dynasty, as its predecessor state, the Xia Dynasty (c. 2070–1600 BC), was in the twilight of its cycle and beginning to lose the Mandate of Heaven. This moral corruption and political upheaval, in keeping with Chinese cosmological belief, was mirrored in a similar upheaval in the heavens, which *Mo Tzu* describes as follows:

> When it came to King Jie of Xia, Heaven gave severe order. Sun and moon did not appear on time. Winter and summer came irregularly. The five grains were dried up to death. Ghosts called in the country, and cranes shrieked for more than ten nights. Heaven then commissioned Tang in the Biao Palace, to receive the great trust that had been given to Xia, as the conduct of Xia fell into great perversity. (Sturgeon 2011)

If true, the climatic events alluded to here would be consistent with the Shiva hypothesis, and may possibly be a recording of an environmental upheaval the likes of which would make the modern-day climate-change threat pale in comparison. Hence its preservation among the foundational myths of so many disparate peoples throughout the region (Pankenier 1998).

To Speak of the Past

The Shooting of the Sun myth is one of a corpus of narratives that take place in the ancient past—a corpus the Bunun refer to as *pali qabasan*, from *pali* meaning "to speak of" and *qabasan* meaning "ancient" or "of the past." It is the era of *Min-pakaliva*, or The Great Chaos: a primordial time of magic and miracles. Then, the word *bunun* was used to refer to human beings in general, and did not come to represent the name of the tribe until Japanese ethnographers established that convention at the end of the nineteenth century (Fang 2016).

In the time of *Min-pakaliva*, man could do more than simply communicate with other beings: metamorphosis of human bodies to animal was possible, just as the baby transmuted into the lizard. Such metamorphoses illustrate the thin barrier separating our existence from that of other creatures, and hence the strength of our kinship with them. Moreover, many animals are considered sacred because of this relationship, and for their aid to the Bunun ancestors during the Great Chaos. This cosmology may be defined by the word "*Deqanin*," which also encompasses temporal, ecological, and celestial meanings. The closest analogy in English is perhaps somewhere between "Supreme Being" and "Heaven," touching upon the eternal and transcendent, yet without the overtly religious connotation. It is, in the words of Huang (2016), "the source of regularity and power of the natural universe."

It was this power, *Deqanin*, who appeared in the form of the moon, and who conferred upon man "*Samu*;" the taboos, norms, and laws governing proper behaviour, how to revere nature and express gratitude to the spirits, as well as knowledge of agriculture and ceremonies. This classifies the Shooting of the Sun as a charter myth, as it validates the ritual, interpersonal and productive practices of Bunun society by providing a creation story with a seasonal ritual function and describing the establishment of laws and taboos. It is through *Samu* that social order is maintained, and human beings can live in sustainable balance in the natural world. Hence the importance of passing down *Samu* to the next generation, through the retelling of myths, through song, and through the observance of ceremonies and other traditional activities. Indeed, Bunun are renowned for having some of the longest, most complex rituals among any of the nations on Taiwan (Kirk 1984).

A grounding in *Samu* and a familiarity with their storied place of residence allows the Bunun, or any conscientious interlocutor, to be receptive to the messages being sent to us from the land and the changing environment, and to seek a cooperative strategy that incorporates the landscape and the variety of creatures in it, in the effort to deal with future iterations of like events. The lesson in this approach is that climate change is not something to be confronted like an enemy, but to be accommodated as a mere shift in the matrix of relationships, finding ways to survive and thrive under a new set of circumstances. What we see as disasters are everyday life experiences, creating new opportunities to make accommodations in our relationship with the natural world, rather than bend it to our will.

Moreover, *Samu* is more than a mere idea, or paying lip service to some vague notion of piety: it is a commitment to structuring one's personal and social life in accordance with a number of calendrical rituals, with between 70 and 130 days per year earmarked as sacred or taboo. Of course, these days were reckoned in accordance with the *sinpatumantuk*, or covenant, made between the Bunun and *Deqanin* as told in this myth, and hence they are determined using the lunar calendar. To keep track of these ceremonial obligations, the Bunun developed a calendar that provided a pictorial representation of each ritual on its proper day.[2] The rituals are specific: in one, women are forbidden to bathe during a lunar month; others mandate when pigs are to be slaughtered, and when millet is to be harvested. Another dictates when the ear-shooting festival (*Mala-Ta-Ngia*), a puberty ritual, is to be held (Fang 2016).

Failure to strictly observe the *Samu* has dire consequences. An infraction of the tiniest sort can bring catastrophic results from the spiritual realm. For example, if certain foods are distributed unevenly at a dinner, this would be a minor *faux pas* in Western cultures, but for the Bunun, it risks calling down illness, drought, a bad hunt, or even death.[3] These magic elements have infused every part of Bunun social life since pre-Christian times, with little distinction between the realm of religion and that of everyday life and social interaction. Indeed, this is equally true of many indigenous societies, just as it was of Western culture before its relatively recent secularization (Bowen 2017; Martin 2005).

CHRISTIAN INFLUENCE

The first Christian missionaries who came to Taiwan did so accompanying the Dutch colonizers, who arrived in 1624. Over the centuries that followed, missionaries discovered that conversion of the Han Chinese was

---

[2] This calendar is often cited as being the first form of writing system on Taiwan.

[3] Relationships are built and sustained based on a gift economy in the Bunun nation, which broadly speaking involves the giving of gifts—a portion of the harvest; a cut of meat from a successful hunt—to others in the community without the expectation of immediate reciprocity. Everyone is a giver and a receiver, which supports the survival of the extended family and community, and helps build strong bonds of trust. Social status is determined by how much one gives, not how much one owns. It is important to note that this symbiotic relationship extends to non-human agents as well, using ritual and sacrifice to confer gifts upon the spirits of the natural world, including the animals of the hunt, and receiving gifts from them as well. In such systems, individuals are more focused on what they can contribute to their society, rather than on what they are owed from society (Hung 2013; Pinchot 1995).

a difficult prospect, due largely to their refusal to give up their practice of ancestor worship, so central is it to their perception of self and place in the universe. In contrast, the island's indigenous peoples proved more willing to embrace Christianity. In 1945, Presbyterian missionaries began to arrive in Bunun villages, followed by Catholic priests 10 years later. Though numbers are declining, there remains a high percentage of indigenous Christian population in Taiwan even today (Fang 2016).

In the process of conversion, there was often a divergence in how the missionary viewed the Bunun connection to their mythic past, and the way the Bunun themselves interpreted Christianity. Missionaries wanted to make a clean break with the past, delineating the pre-Christian era from the Christian present. Meanwhile, the Bunun sought to view the Bible through the lens of their existing *weltanschauung*, and in fact were successful in deriving an elegant syncretism between the old worldview and the new. They found several parallels between their cosmology and the Christian stories, the most obvious of which is that both include a take on The Great Flood. The myth of shooting the sun likewise contains several points of remarkable correspondence (ibid.).

For one thing, the myth is similar to the expulsion from paradise, particularly in how *Deqanin* tells the man that he will thenceforth have to rise early to toil in the fields, echoing God's punishment of Adam after he and Eve ate the fruit from the tree of the knowledge of good and evil, to wit Adam would now have to work hard to derive food from the earth. Another similarity is in how the man was tasked with bringing *Samu* to the Bunun people and ensuring they abide by the new rituals and taboos. This marks the man as a culture hero not unlike the lawgiver Moses who descended from Mount Sinai with God's laws. Thus, the Ten Commandments were essentially viewed as the *Samu* of the Israelites. Nor are these parallels restricted to the Old Testament. Early Bunun converts noticed that there was an equivalency between *Deqanin*, appearing as the moon—the sun whose wounding ushered in a cosmic transformation—and the figure of Christ, whose suffering on the cross brought mankind a New Covenant and whose sacrifice marked the beginning of a new era of mankind. Moreover, in a general sense, both cosmologies—traditional Bunun and Christian—are inherently cognizant and respectful of the power of words as the primary creative force to bring reality into being. Ten iterations of the phrase "God said" can be found in the book of Genesis to describe the divine act of creation through the mere utterance of words. Likewise, it was the words of *Deqanin* that brought into existence the rituals that formed a mutually reinforcing covenant between the Bunun people and the universe, just as that same power is invoked with every telling of the myth, continually re-creating the culture through the speaking of it. It is not surprising therefore that many Bunun came to recognize in the Christian God the same supernatural spirits of the religion their ancient ancestors believed in (Martin 2005; Fang 2016).

PSYCHOLOGICAL SIGNIFICANCE

Chaos is how we interpret the outside world—especially the natural world: dangerous, unpredictable, and unbowed by our attempts to control or even understand it. In contrast, society represents order, control, and the rules and laws that govern human-to-human interaction. No society can survive a totality of one and an absence of the other: rather, a balance must be struck, and so it is the task of the hero to venture out into the chaotic world, to risk death in order to undertake an adventure, learning and being changed by the chaos, and to bring home this new knowledge as a boon to the community. This myth therefore echoes Marduk defeating Tiamat and creating the world (civilization) from her remains; Indra combating Vritra, the demon of drought; Ra versus Apophis, one of whose titles was "Lord of Chaos;" and countless other examples. The three stages of the hero's journey are clearly in evidence in this tale—Separation (leaving the regular world for a magical realm), Initiation (risking death to defeat supernatural forces), and Return (coming home again with tools to heal the community) (Campbell 2008).

The victory over chaos in worldwide mythologies often marks the beginning of a new civilizational era, bringing in the new gods of culture, and sidelining or outright defeating the old gods of nature. It often accompanies the shift from subsistence-lifestyle forager societies—which were at the mercy of the caprice of the mysterious and unknowable natural world—to a more stable form of food production through farming and herding, which for the first time availed a surplus, and hence the ability to engage in ceremonial redistribution to mitigate inequality. Societies with this heightened level of complexity demanded stronger social cohesion and new rules governing human–human and human–nature interaction. The Bunun myth perfectly fits this pattern, and serves as an example of this phenomenon: the creation of the moon, which yet is a continuation in spirit of the wounded sun from which it emerged, and the conferring of new laws and rituals (the lunar cycle) to govern correct living and prosperity in the new Bunun post-agricultural revolution society (Christ 1987; Scheidel 2018).

What is interesting about this Bunun myth is that, unlike most other myths of this sort, there is a humility inherent in the narrative. Unlike Hou Yi's dominance over nature,[4] Marduk's destruction of the primordial goddess Tiamat; Apollo conquering the earth-spirit Python; and countless other examples, the Bunun hero, blinded by his obsession with revenge, is revealed as the true antagonist of the story, or at least the agent of his own misfortune (Fang 2016).

It is therefore more accurate to conceptualize this myth as descriptive of a syncretisation of beliefs rather than a supplantation, wherein notwithstanding the shift to farming and herding—a drawn-out process that was not always linear—there was a continuation of the hunter-gatherer culture,[5] and both the old gods and the new gods were necessary in the universe. This syncretisation is beautifully illustrated by the transmutation of the wounded sun into the moon, by use of the man's arrow deployed in anger, and the man's cloth, offered in compassion and contrition (Scheidel 2018).

The wounded sun opens the man's eyes to his own agency, and in this sense man (society) is entering adulthood—his eyes opening for the first time to his responsibility for healing his childhood wounds and controlling the neuroses linked with adolescence. Man, in the Bunun story, is not merely the hapless plaything of the gods whose enmities and alliances dictate events on the earthly realm, but a player with the power

---

[4] A prevalent Chinese interpretation of that culture's version of the same myth draws the exact opposite lesson from the one drawn in the Bunun myth. In his victory against the nine suns, Hou Yi did not make a pact with the forces of the universe and of the natural world around him: rather, his achievement is interpreted as representing man's dominance over nature, and civilization's taming of mother earth, through the power of collective farming and the people's capacity to resist even drought in their agricultural production goals, and in their lives. This is consistent with the narrative being pushed by the Chinese Communist Party, wherein the Maoist revolution was an attempt to "conquer nature" as much as it was to conquer the Nationalists. Thus, the lessons learned in China from the superfluous suns myth are about resistance, tenacity, and willpower: Hou Yi achieved man's dominance over the deities, and over nature itself (Guo et al. 2020; Shapiro 2001).

[5] Even today, the Bunun are fighting to retain their hunting rights, so central is it to their identity.

to influence events in his relationship with the supernatural, and therefore possessed of a responsibility to do so with wisdom, humility, and compassion.

Progressive Tension

In the Bunun myth, it is the moon that represents a masculine energy and the sun that exhibits the feminine. This may seem like an inversion of traditional wisdom, but in fact there is ample precedent. Anthropologists have long observed that cultures, both archaic and modern, imbue these two celestial bodies—sun and moon—with divine properties, and their dance through the sky represents the basic tension that moves a healthy society forward. The opposing forces creating this progressive tension (in both society and psychology) are often saddled with the labels "masculine" and "feminine." The solar-masculine and lunar-feminine paradigm has dominated in those belief systems that are derived from the Greek and Roman Judeo-Christian traditions, and this has become received wisdom in Western society. This was not always the case, however: many archaic mythological systems identified masculine qualities with the moon, and considered the sun exemplary of feminine energy. These earlier, neolithic mythologies held the female above the male. It was only with the advent of agriculture, the establishment of sedentary lifestyles, and eventually the beginnings of city-states, that this paradigm was inverted, first in Mesopotamia, to a male-oriented pantheon, culminating in the Abrahamic monotheistic religions.

Joseph Campbell is often excoriated for seeming to support this assignment of gender labels to solar and lunar values, which served to stigmatize the feminine and allow patriarchal systems to supplant belief systems that venerated the mysterious and unknowable natural word. In fact, he was describing this process of the manipulation of mythology by power holders as a means to support their new social order, and lamented this suppression of the lunar-masculine and solar-feminine in favour of the narrow paradigm we employ in the Western world today: that of the lunar-feminine and solar-masculine (Campbell 2018; McCrickard 1991; Teich 2015).

The mythological record abounds with examples of lunar-masculine and/or solar-feminine deities: in Germanic mythology, Sól is the sun goddess and sister to Máni, who embodies the moon. Both will be destroyed in Ragnarök, it is told. Akycha is the sun goddess of Inuit

mythology who races across the sky as she is being chased by her inces-
tuous brother, moon. Tsukuyomi-no-Mikoto is the moon god of Shinto
mythology in Japan. Even the great Thoth, the scribe and adviser to the
Egyptian gods, was originally a moon deity before he became associated with magic and knowledge. No, there is nothing universal about
the concept of a feminine moon and masculine sun, and the Bunun myth
can be counted among these other examples that stand as testament to
this fact.

The Sun/Moon Dyad

The sun in the Bunun myth represents chaos, and the mysterious forces
of nature that puts man at its mercy. The moon, in contrast, represents
ritual; order; agriculture: in short, civilization. Yet the tale also provides
us with a mixing of properties: the moon comes from the sun, and has
solar qualities; whereas the sun always had within it the potentiality of the
moon. It is therefore unlike the Western phenomenon of solarisation—of
victory of the one over the other—and more of an amalgamation. It is, in
short, a fine example of synthesis of the sort Campbell identified a need
for, wherein the masculine was incomplete until it embraced its feminine
energies.

We see this too not just in the sun-to-moon transformation, but in
the epiphany experienced by the hero. Only through humility and acceptance of his agency and responsibility for himself and his world was he
able to emerge into the hero he was meant to be, and thus to bring
home boons to his community. It is this worldview that is the greatest
lesson that Western culture can take away from the Bunun experience as
today's scientists seek to incorporate TEK into their efforts to address
climate change, if indeed they are serious about that goal. There are
several harms that have ensued from Western culture's zero-sum view of
solar and lunar principles, and the approach to climate change is but one
of them. Nevertheless, it is an important one.

Lessons Learned

What can we learn from the Bunun and other indigenous peoples? Much
of the scientific research on indigenous oral histories focuses on lever-
aging the collective memory to serve as an extremely long record of plant
species change, faunal migration and habitat shifts, and other clues to

describing what long-term climate-change patterns may have taken place since the ancient past. As discussed in this chapter, however, there may be deeper lessons to be learned (Robbins 2018).

There is scientific precedence attesting to the success achieved by certain preindustrial societies and communities in dealing with climate upheavals in the past through resilience and adaptation of the sort evidenced by the Bunun (see Degroot 2018). In contrast, the modern world largely seeks, through its climate-change policies, to stop global warming and revert to a more comfortable and familiar *status quo ante*, precisely so that we don't have to change our lifestyles. Several scholars have offered ways in which our global culture can, in the model of the Bunun, adopt a better perspective from which to view our place in the world and to proceed towards better solutions.

Rachael A. Vaughan (2020) examines how the hero archetype renders the Western ego unequal to the task of properly addressing climate change because of the defensiveness and unconscious patterns it engenders, proposing adoption of an "archetype of the initiate" to better navigate our uncertain future. She points out that our cultural predisposition is to solve problems by slaying dragons, and that we are approaching climate change as an external threat attacking us (we being innocent victims) from the outside. Thus, our focus has been on finding a villain in this narrative, and attacking it. As if to illustrate this very point, US Special Presidential Envoy for Climate John Kerry said at a June 28, 2021, press conference that the international community needs to assume a "wartime mentality" to tackle climate change; a mentality in which presumably the changing climate plays the role of the enemy—an enemy that needs to be vanquished. Vaughan identifies, as a consequence of this mindset, the naming of various enemies and launching crusades to slay them. The examples she provides include efforts to sue the heads of oil corporations for causing emissions, pointing out that the slaying of this particular dragon may make us feel good for a while, but it will not lead to any change in consumers' fossil-fuel consumption habits, nor will it lead to an acknowledgement of our own culpability (Jessop 2021; Vaughan 2020).

Citing Jung (1989), Vaughan points to the dangers of not maturing past the psychological stage of engaging in heroic adventures, and putting off the goal of individuation. Psychological maturity past the hero is the archetype of the Initiate, and is marked by a wilful cession of control and submission to the uncertainties of life, wherein the only reliable truth is change. She suggests that, as a civilization, we must adopt this "archetype of the initiate" to replace the hero archetype that has brought us to this juncture and which may no longer be appropriate for solving this and other problems of the twenty-first century (ibid.; Henderson 2005).

Shedding ourselves of the hero archetype would necessitate new values and new goals, and the voluntary adoption of a life based on humility, resourcefulness, and sharing. First, however, there would have to be a collective acknowledgement of mankind's shared culpability for the climate crisis in which we are embroiled, and the development of a renewed appreciation for the biosphere of which we are a mere part. In other words: you aren't *in* traffic; you *are* traffic. This would lead to the realization that continuance of life, rather than our own gratification, is a superior goal. Leadership in the archetype of the initiate is humble, attentive, and inclusive, and only through a "conscious dialogue with the Self, the world, and the world's others in all their forms" will we as a civilization be able to navigate a way through the climate crisis (Laling, n.d.; Vaughan 2020).

Echoing Vaughan's call for an "archetype of the initiate," Sam Adelman (2015) believes that the global ecological crisis was exacerbated, if not outright caused, by Western civilization's reliance on epistemologies of "mastery," through which we view it as our right to hold dominion over the natural world. Adelman casts doubt on the notion that our Western-derived science, legal frameworks, political structures, and culture are even capable of handling such an immense problem as climate change. He is in agreement with Vaughan that the answer to climate change and biodiversity loss lies not in more developmentalism, extractivism, neoliberalism, and other ideologies that rely on growth and technology to return the climate to some idealized state. In his words, "Seeking ways to engineer the climate suggests that human beings have learned nothing and forgotten nothing" (Adelman 2015). Rather, the answer lies in shunning anthropocentrism and the view of progress as defined by Enlightenment rationality in favour of a new conception of modernity in which humanity is no longer at odds with nature. Put succinctly; we are not *on* the mountain, we are *of* the mountain—or we should be (Yumin, n.d.; Adelman 2015; León 2012).

What does that mean in practical terms? According to Hamilton (2013), first and foremost it would necessitate the abandonment of the large-scale geoengineering mega-projects aimed at addressing the climate issue, including technologies for carbon capture and sequestration. These are "technologies of hubris," according to Harvey (2003), as they are contingent upon expertise that is organized hierarchically, with centralized decision-making, and are overly dependent on the "cult of the expert." Gigantic engineering endeavours will not get us out of the mess that we are only in because of similar industrial-scale projects, built using similar technologies of hubris. Instead, what are needed are what Jasanoff (2005) terms technologies of humility, which "compel us to reflect on the sources of ambiguity [and to] think harder about how to reframe problems so that their ethical dimensions are brought to light" (Jasanoff 2005).

The goal of instituting a more sustainable and equitable future by transforming the entire socio-economic system would necessitate a new paradigm that downsizes the scale of economic processes so as not to exceed the limits of the planet's biophysical carrying capacity. There is a lack of a theoretical framework that might serve as a roadmap for how such a transition could even be attempted, according to Ruder and Sanniti (2019). In their view, even the promising field of ecological economics is too rooted in androcentric disciplines to serve as the source of theoretical analysis, much less practical applications. To assess this deficiency, they employ a gender analysis to examine the patriarchal roots of our current growth-oriented capitalist paradigm. Though ecological economics explores post-growth systems for distribution of material, spiritual, and financial assets, the transitions predicated thereupon would inevitably result in limiting standards of living and replacing fossil fuels as the primary source of energy in the economic system with raw, human labour (Ruder and Sanniti 2019).

If this transition described by Ruder and Sanniti materializes, it would essentially create a reversal of the process described by writers such as Nikiforuk (2012), Ridley (2010), and Mouhot (2011), whereby the harnessing of fossil fuels replaced slaves as the source of raw energy that has powered the global economy since slavery first appeared in Mesopotamia around 9,000 years ago. This is not to say that a wholesale reversal of this progress will inevitably result, but certainly human labour would be commodified and the cheapest sources thereof prioritized. In the words of Andrew Nikiforuk (2012), "We feel entitled to surplus energy and rationalize inequality, even barbarity, to get it."

Conclusion

In sum, the true lesson to be learned from the Bunun—and indeed from any indigenous peoples' TEK—is not about melting Arctic ice, protecting fish stocks, and controlling wildfires (Robbins 2018). These things are important, to be sure. But it is a deeper lesson, about Western society's focus on Enlightenment values, and the need to incorporate the lunar-masculine and solar-feminine in a new worldview characterized by humility and connectedness to the natural world and each other. Humanist and religious discourse has been marginalized in the study of global ecosystems. As proposed by Timothy Leduc (2010), a cultural or spiritual revision must take place if industrialized societies are going to be able to chart a sustainable path forward for the planet. This revision would necessitate a transformation in how interdisciplinary research conceptualizes climate change, and hence how best to respond to it. In the words of conservationist Paul Shepard (1998), the indigenous approach can help "carry us beyond ourselves, pursuing the nature of thought as the thought of nature" (Shepard, 37).

For all his highly publicized faults, Campbell (2008) was acutely aware of this, writing: "Nor can the great world religions, as at present understood, meet the requirement. The universal triumph of the secular state has thrown all religious organizations into such a definitely secondary, and finally ineffectual, position that religious pantomime is hardly more today than a sanctimonious exercise for Sunday morning, whereas business ethics and patriotism stand for the remainder of the week."

Although there have been many studies of indigenous land-use practices and how information obtained from TEK can be used to buttress governments' climate-change policies, we hope to have shown in this chapter that information, while important, is not enough, and that wisdom should be the goal. This has been done by presenting the case study of the Bunun myth of the Shooting of the Sun, the role it plays in Bunun society, and the pivotal position it has in the formation of the Bunun worldview characterized by humility and a mindfulness of the relationships that provide meaning in life—relationships with the natural world, with the spirits in it, and with each other. By analysing this myth, a clear line becomes evident from indigenous belief systems to their experiential relationships with the natural world, which places the human experience in balance with the ecosystem in a far healthier way than is conceived of by our Western social order.

Campbell (2008) believed a transmutation of the whole social order was necessary so that each individual's vitalizing agency and consciousness of their place in the universe would be evident "in every detail and act of secular life" (Campbell, 360). This was not an imminent possibility in Campbell's day, and we are scarcely closer now. While there has been, in the West, a widespread questioning of the wisdom of the Enlightenment and reassessment of the virtues of Western culture, there are as yet no philosophical foundations or social movements that even attempt to remedy these faults, most being content to tear Western civilization down without sparing a thought for what should be built up in its place. Indeed, there seems to have been a recapitulation of the rationalist preoccupation of the Enlightenment and its denial of esoteric wisdom, to the point that the mantra of 2021 has been "follow the science." Only science, we are told, will save us from the global disaster in which we find ourselves. Only, science, we are learning, may well be what got us here in the first place.

REFERENCES

Adelman, S. (2015). Epistemologies of mastery (pp. 9–27). In A. Grear & L. J. Kotzé (Eds.), *Research handbook on human rights and the environment*. Edward Elgar Publishing.

Allan, Sarah. (2010, February). T'ien and Shang Ti in Pre-Han China. *Acta Asiatica 98*, 1–18.

Anderson, B. (2006). *Imagined communities: Reflections on the origin and spread of nationalism*. Verso Books.

Berkes, F. (1985). Fisherman and the tragedy of the commons. *Environmental Conservation 12*, 199–206.

Berkes, F. (2012). *Sacred ecology*. Routledge.

Bowen, J. R. (2017). *Religions in practice: An approach to the anthropology of religion*. Routledge.

Campbell, J. (2004). *Pathways to bliss: Mythology and personal transformation* (Vol. 16). New World Library.

Campbell, J. (2008). *The hero with a thousand faces* (Vol. 17). New World Library.

Campbell, J. (2018). *Oriental mythology* (Vol. 2). Joseph Campbell Foundation.

Christ, C. P. (1987). Toward a paradigm shift in the academy and religious studies. The articulation of gender as an analytical category (pp. 53–76). In C. Farnham (Ed.), *The impact of feminist research in the academy*. Indiana University Press.

Costa-Pierce, B. A. (1987). Aquaculture in ancient Hawaii. *Bioscience 37*, 320–331.

DayDay News. (2019, November 11). *Is Houyi shooting the sun just a myth? NASA scientist: No! That might be true!* https://daydaynews.cc/en/history/215822.html

Degroot, D. (2018). *The frigid golden age: Climate change, the little ice age, and the Dutch Republic, 1560–1720.* Cambridge University Press.

Du, S., & Winkler, R. J. (2003). *Bunong zu: Yu yue liang de yue ding* [Rendezvous with the moon: Stories from the Bunun tribe]. Taibei Shi: Xin zi ran zhu yi gu fen you xian gong si.

Fang, C. W. (2016). *Transforming tradition in Eastern Taiwan: Bunun incorporation of Christianity in their spirit relationships* [Doctoral dissertation, The Australian National University]. Researchgate.

Gare, A. (2001). Narratives and the ethics and politics of environmentalism: The transformative power of stories. *Theory and Science 2*(1).

Guo, R., Zhang, Y., & Shen, H. (2020). Psychological significance of Chinese myth of Yi: Signing and symbolizing, cultural and personal. *Culture & Psychology 26*(2), 234–252.

Hamilton, C. (2013). *Earthmasters: The dawn of the age of climate engineering.* Yale University Press.

Harvey, D. (2003). The fetish of technology: Causes and consequences. *Macalester International 13*(1), 3–30.

Henderson, J. L. (2005). *Thresholds of initiation.* Chiron Publications.

Ho, Ting-jui. (1971). *A comparative study of myths and legends of Formosan aborigines.* Taipei, ROC: The Orient Cultural Service.

Huang, Hsinya. (2016). (W)ri(gh)ting climate change in Neqou Soqluman's work (pp. 17–28). In C.-j. Chang & S. Slovic (Eds.), *Ecocriticism in Taiwan: Identity, environment, and the arts.* Lexington Books.

Hung, H. R. (2013). *Cultural survival, Indigenous knowledge, and connectedness: A comparison of case studies in Taiwan and Australia* [Doctoral dissertation, Macquarie University, Sydney, Australia]. researchonline.mq.edu.au.

Jasanoff, S. (2005). Technologies of humility: Citizen participation in governing science (pp. 370–389). In A. Bogner & H. Torgersen (Eds.), *Wozu Experten?: Ambivalenzen der Beziehung von Wissenschaft und Politik.* Wiesbaden: VS Verlag für Sozialwissenschaften.

Jessop, Simon. (2021, June 28). US envoy Kerry says world needs a 'wartime mentality' over climate. *Reuters.* https://www.reuters.com/business/environment/us-envoy-kerry-says-world-needs-wartime-mentality-over-climate-2021-06-28/

Jung, C. G. (1989). Memories, dreams, reflections (A. Jaffé, Ed.) (R. Winston & C. Winston, Trans.). New York: Vintage. (Original work published 1961).

Karalekas, D. [dkaralekas] (2013, May 25). *Rendezvous with the moon: A Bunun myth* [Video]. YouTube. https://youtu.be/IrCnddinc9c

Kirk, G. S. (1984). On defining myths (pp. 53–61). In A. Dundes (Ed.), *Sacred narrative: Readings in the theory of myth*. University of California Press.

Kuhlmann, D. (2019). Writing in my voice: Four modalities of myth-writing in Taiwanese Indigenous Sinophone literature. *Ex-position 42*, 29–51.

Leduc, Timothy. (2010). Climate research, interdisciplinarity, and the spirit of multi-scalar thought (pp. 119–144). In S. Bergmann & D. Gerten (Eds.), *Religion and dangerous environmental change: Transdisciplinary perspectives on the ethics of climate and sustainability* (Vol. 2). Münster: LIT Verlag.

León, M. (2012). Economic redefinitions toward buen vivir in Ecuador: A feminist approach. *Feminist perspectives towards transforming economic power: Topic, 2*.

Martin, J. F., Roy, E. D., Diemont, S. A., & Ferguson, B. G. (2010). Traditional Ecological Knowledge (TEK): Ideas, inspiration, and designs for ecological engineering. *Ecological Engineering 36*(7), 839–849.

Martin, Steven Andrew. (2005). Ethnohistorical perspectives of the Bunun: A case study of Laipunuk, Taiwan [Master's Thesis, National Chengchi University]. NCCU Institutional Repository.

McCrickard, J. E. (1991). Born-again moon: Fundamentalism in Christianity and the feminist spirituality movement. *Feminist Review 37*(1), 59–67.

McIntosh, C. (2012). *The rose cross and the age of reason: Eighteenth-century Rosicrucianism in Central Europe and its relationship to the enlightenment*. Suny Press.

McNeill, L. S. (2013). *Folklore rules: A fun, quick, and useful introduction to the field of academic folklore studies*. University Press of Colorado.

Metevelis, P. (2005). The Dog Star and the Multiple Suns motif: An Asian contribution to European mythology. *Asian Folklore Studies 64*(1), 133–137. http://www.jstor.org/stable/30030361

Miciq, Rangi. (2021, June 25). Personal communication [Personal interview].

Mindlin, B. (2002). O fogo e as chamas dos mitos. Estudos Avançados [The fire and the flames of myths]. *Estudos Avançados 16*(44), 149–169.

Mouhot, J. F. (2011). Past connections and present similarities in slave ownership and fossil fuel usage. *Climatic Change 105*(1), 329–355.

Müller, F. M. (1856). *Comparative mythology*. Routledge.

Napier, W. M., & Clube, S. V. M. (1979). A theory of terrestrial catastrophism. *Nature 282*(5738), 455–459.

Niamir-Fuller, M. (1998). The resilience of pastoral herding in Sahelian Africa (pp. 250–284). In F. Berkes & C. Folke (Eds.), *Linking social and ecological systems: Management practices and social mechanisms for building resilience*. Cambridge: Cambridge University Press.

Nikiforuk, A. (2012). *The energy of slaves: Oil and the new servitude*. Greystone Books.

Pankenier, D. W. (1998). Heaven-sent: Understanding cosmic disaster in Chinese myth and history (p. 187). In B. J. Peiser, T. Palmer, & M. E. Bailey (Eds.), *Natural catastrophes during bronze age civilisations: Archaeological, geological, astronomical and cultural perspectives*. Oxford: Archaeopress.

Pearce, F. (1993, September 11–12). Living in harmony with forests. *New Scientist 23*.

Pinchot, Gifford. (1995). The gift economy. *Context 41.* https://www.context.org/iclib/ic41/pinchotg/

Pinker, S. (2018). *Enlightenment now: The case for reason, science, humanism, and progress*. Penguin.

Quintana-Ascencio, P. F., Gonzalez-Espinosa, M., Ramirez-Marcial, N., Dominguez-Vazquez, G., & Martinez-Ico, M. (1996). Soil seed bank and regeneration of tropical rain forest from milpa fields at the Selva Lacandona, Chiapas, Mexico. *Biotropica 28*, 192–209.

Ridley, Matt. (2010). *The rational optimist: How prosperity evolves*. New York: Harper

Rinkevich, S., Greenwood, K., & Leonetti, C. (2011). Traditional ecological knowledge for application by service scientists. *US Fish & Wildlife Service*.

Robbins, Jim. (2018, April 26). Native knowledge: What ecologists are learning from indigenous people. *Yale Environment 360*. https://e360.yale.edu/features/native-knowledge-what-ecologists-are-learning-from-indigenous-people

Ruder, S. L., & Sanniti, S. R. (2019). Transcending the learned ignorance of predatory ontologies: A research agenda for an ecofeminist-informed ecological economics. *Sustainability 11*(5), 1479.

Scheidel, W. (2018). *The great leveler: Violence and the history of inequality from the stone age to the twenty-first century*. Princeton University Press.

Segal, R. A. (2016). Friedrich Max Müller on religion and myth. *Publications of the English Goethe Society 85*(2–3), 135–144.

Shapiro, J. (2001). *Mao's war against nature: Politics and the environment in revolutionary China*. Cambridge University Press.

Shepard, P. (1998). *Thinking animals: Animals and the development of human intelligence*. University of Georgia Press.

Sturgeon, Donald. (Ed.). 2011. *Chinese text project*. http://ctext.org

Sweeney, J. F. (2013). *Fifth stone of the sun and the Qi Men Dun Jia model*.

Teich, H. (2015). The twin heroes: Campbell's solar/lunar vision of the masculine (pp. 87–103). In K. L. Golden (Ed.), *Uses of comparative mythology (RLE myth): Essays on the work of Joseph Campbell*. Routledge.

Thomas, W. H. (2003). One last chance: Tapping indigenous knowledge to produce sustainable conservation policies. *Futures 35*, 989–998.

Vaughan, R. (2020). The hero versus the initiate: The western ego faced with climate chaos. *Journal of Jungian Scholarly Studies 15*(1), 48–62.

Whyte, K. P. (2013). On the role of traditional ecological knowledge as a collaborative concept: A philosophical study. *Ecological Process 2*, 7. https://doi.org/10.1186/2192-1709-2-7

Yumin, Laling. (2016, August). Personal communication [Personal interview].

Zhai, P., Pirani, A., Connors, S. L., Péan, C., Berger, S., Caud, N., ... Goldfarb, L. (2021). *Climate change 2021: The physical science basis*. Contribution of Working Group I to the Sixth Assessment Report of the Intergovernmental Panel on Climate Change.

North American Native Literature and Environment: Perspectives on the Native Challenges and Dispossession

Zakia Firdaus and Amar Wayal

On October 12, 1492, Columbus arrived at one of the Bahamian islands in the West Indies. Believing that he had reached the outer boundaries of Asia, accordingly, he named these islanders the "Los Indios, the inhabitants, Indians" (Marder 2005: 1). As a result, October 12, 1492, remains the official date of the first recorded discovery of America and the first documented interactions with the people of the New World. Columbus used the term "Indios" of which variations from the late 1400s remained the same in order to make the inhabitants and Indians Christian by portraying them savage through European eyes.

Z. Firdaus
Center for Comparative Literature and Translation Studies, Central University of Gujarat, Gandhinagar, India

A. Wayal (✉)
Department of English, SRM Institute of Science and Technology (Deemed to be University), Tiruchirappalli, India
e-mail: amarwayal8@gmail.com

© The Author(s), under exclusive license to Springer Nature Switzerland AG 2022
N. J. P. Alsford (ed.), *Pacific Voices and Climate Change*,
https://doi.org/10.1007/978-3-030-98460-1_8

Columbus's arrival marked the opening of cultural injustice, coloniza-tion, violent usurpations of native lands, and brutalization of native inhab-itants. The subsequent explorers' occupied shores as varied as Columbus was discovered the huge cultural variations. As Anthony Pagden states, the explorer, the discoverer, the settler, the refugee, the missionary, and the colonist came to America with broken hopes, conflicting aspirations, and different goals to transform unfamiliar cultures of which they had no prior understanding of culture and language. To the continuation of these clashes, the resulting atrocities were set up with grave difficulties, wars, famines, epidemics, and general uncertainty (Carpenter 1994: 2–3).

Between the early 1500s and the late 1600s, five waves of Europeans landed in what is now the continental United States:

> Spanish into the Floridas and the southeast from South America and the Caribbean islands, Spanish into Texas, the desert southwest, and California from Mexico, English with landings along the Atlantic seaboard, French into the Great Lakes region and Mississippi Valley from Canada, French, and Spanish with landings along the Gulf Coast. (Santoro 2009: 5)

Throughout the North American continent, early European explorers, traders, and settlers from Spain, France, and Great Britain encountered Native tribes. For the Europeans, it was a whole new universe. For thou-sands of years, it had been the home of the indigenous people. Hunters and traders took over from the early explorers. Missionaries and settlers took their place as a result of their actions. The Spanish and French gave way to the English during the next hundred years, and the English gave way to those who adopted the name American as their own. These new Americans moved west, pushing the Native Americans even further ahead of them until they finally pushed beyond the Mississippi, except for a small minority remained on reserves.

Lands allocated to immigrant Indians consisted of "tribal reservations carved from the territory and, for the most part, located along the eastern border of what later became Nebraska, Kanas, and Oklahoma" (Chalfant 2002: 4). Their removal also caused friction with the Plains and South-west Indians, who, like those on the Pacific Coast, were driven in different directions by an influx of white traders, farmers, and miners. The Euro-pean missionaries drove the natives out of their homeland. Because the United States government decided to locate Indian reservations within state borders, those states grumbled about being expected to provide. The

United States government established Indian reservations to subjugate native people, with disastrous results and long-term consequences.

Environment plays a pivotal place in North American literary landscape. As Paula Gunn Allen has written explicitly, "We are land-is the fundamental idea embedded in Native American life and culture" (Porter 2012: 65). It constitutes their notion of the self. Consequently, there has been a conscious effort by writers to raise concerns on environmental degradation and sustainability. Large corpus of native writing is engaged in ecocriticism and issues on protection/preservation of natural resources. The reservation is one of the key areas deliberated by writers to engage the larger debate on how their space/region is violated by the dominant order. For instance, the American Southwest has been seen a site of atomic power block in the shape of uranium mining, atomic power development and atomic testing program. In fact, that area became so devastated for human habitat that it was declared as "National Sacrifice Area" under Nixon administration. Along with this, there has been other regions around reservations areas which have been used as space for industry and other power projects. The present paper strives to study the perspectives on the Native challenges and dispossession from the praxis of the environment and its nature. Environmental challenges of North American Native culture are a critical literary expression in examining the larger problems of Native people and the natural world around them and subverting the distinctions between the categories of Native and Non-Native people, the protectors of nature, and exploiters of nature. North American Native literature reiterates Native challenges and dispossession to sustain the potential witness to an integrated vision of environmental activism with their own homeland. This paper analyzes Leslie Marmon Silko's *Ceremony*, Scott Momaday's *House Made of Dawn*, Louise Erdrich's *Tracks*, and Joseph Bruchac's *Long River* from the position of the capitalist development, which denies the needs of Native people and their land. These novels examine cultural and literary representations of the environment, which has been colonized by the mainstream culture. These authors make a skilful narration of the Native oral tradition to express the reality of Native experience by connecting the land to their Native cultural heritage. Perspectives on the Native challenges and dispossession in their works not only comment on dominant spaces but also evolve Native spaces into nature. Leslie Marmon Silko, Scott Momaday, Louise Erdrich, and Joseph Bruchac belong to their respective tribal positions and are engaged in a narrative which not only criticizes colonialism but also incorporates myth, history, and contemporary issues in order to posit the narrative on survival. This also runs parallel to their idea of communal living where the concept of one is for all and all are for one is projected. Incidentally, contemporary writers are making a stronger presence by critiquing various projects of the government and other agencies, which have impacted the native sense of nature and land.

Environmental Degradation on the Reservation

Reservation policies in the nineteenth century have been a tool for reducing the continental expanse of Indian lands in the path of European settlement and transforming Indians into idealized farmers. These policies "contributed to the alienation of Indian reservation lands-a diminishment of land, resources, and biotic diversity that relegated Indians to the political and economic periphery of American society" (Lewis 1995: 423). The location of reserve lands across North America has long been considered a crucial notion in Native Americans' view of nature and self in isolated settings. As Paula Allen Gunn has spoken of, earth is the mind "of the people as we the mind of the earth. It is not a matter of being close to nature. The relationship is, in a very real sense, the same as ourself, and it is this primary point that is made in the fiction and poetry of the Native American writers of the Southwest" (Porter 2012: 65–66). Their Reservation land allows them to re-center their self-awareness and have a direct relationship with the environment. The land and sacred places that bordered and formed their world helped Native people define themselves.

Land usage in agriculture and grazing had drastically changed the face of native reservations by the turn of the twentieth century. This fact indicates that the challenges confronting Indians and their environment are defined by land exploitation, changing Indian requirements, attitudes, and religious demands, as well as evolving Indian wants, attitudes, and religious needs. The territory that Euroamerican claimed as wilderness had been modified by Indian cultivation, irrigation, and field distribution networks. As Vine Deloria says in this context that the "wilderness is contained the gulf between the understandings of the two cultures. Indians do not see the natural world as a wilderness. In contrast, European and Euroamerican see a big difference between lands they have settled and lands they left alone" (Deloria 1992: 281). Deloria's argument goes against the deep ecology political viewpoint of preserving ecosystems by forbidding or restricting human presence. The implicit assumption here is that native lifestyles do not harm the landscape in the same way that contemporary urban settlements and resource extraction do.

This environmental degradation causes tensions between native peoples and non-native governments over natural resources needed to support tribal life. They mediate between their members' divergent expectations and the BIA's industrial attitude to balance reservation land for economic needs and cultural preservation. State and federal officials tried to restrict Indian off-reservation subsistence rights, including pay work in the same industries, the loss of fishing locations, and natural resource extraction, at the demand of commercial interests. There "are no environmental rules and regulations...no one cares about the people who live" (Lewis 1995: 431) on the reservation. This law appears to oppose some of the most environmentally damaging forms of exploitation, such as mine and drilling sites, roads and machinery, tailing piles, and settling ponds, all linked to tribal land, water, air, health, and lifestyles. Mining and oil and gas development have scarred thousands of acres with minimal protection for native people, despite efforts by Indian activists to balance resource use and safety. Native people suffer environmental concerns related to daily living conditions, changing diets, and urban growth due to these negative manifestations. Likewise, diet and activity changes—"a result of changing subsistence patterns and environments-contributed to an explosion of dietary-related illnesses like diabetes, vitamin and mineral deficiencies, cirrhosis, obesity, gallbladder disease, hypertension, and heart disease" (Lewis 1995: 437).

Native people who have used political activity to fight multinational corporations and the government to stop environmentally disastrous projects on their lands have been criminalized and jailed to quiet their claims. Both reservation cities and border towns are seeing fast growth, which is causing environmental issues. Indian communities deal with zoning and housing concerns, solid waste disposal issues, and municipal and agricultural wastewater treatment. The question of power and control over resources is beginning to shift as Indigenous peoples continue to challenge the power structure of multinational corporations and the state, asserting their sovereignty rights as native people to control natural resources within their territories according to treaties. In discussing social control and land issues:

> land has always been the issue of greatest importance to politics and economics in this country. Those who control the land are those who control

the resources within and upon it. No matter what the resource issue at
hand is, social control and all the other aggregate components of power
are fundamentally interrelated. (Robyn 2002: 209)

In ways that European immigrants do not, Native Americans appreciate
and revere the land, the environment, and the human interconnectedness
to that ecology. Many Native American communities' ecological farming
and hunting traditions have supported their millions of people for thousands of years without damaging the land or its animal and vegetable
species. Native Americans lived in harmony with their surroundings, but
they also aggressively exploited environmental forces to satisfy their physical requirements. The employment of tools to increase human impact on
the natural environment is referred to as technological advances. Native
Americans relied heavily on fire to manage forest and grassland resources
and numerous hunting implements, agricultural implements, irrigation,
and other water management methods for agriculture and astronomy
tools.

This is what Scott Momaday refers to this knowledge and understanding of the natural world "reciprocal appropriation by which he
means that human beings invest themselves in the landscape, and at the
same time incorporate the landscape into [their] own most fundamental
experience" (Schweninger 1993: 48). The knowledge and understanding
of the natural world are meant to be a thread woven in the wake of
the environment that plays a large role in the works of Leslie Marmon
Silko, Louise Erdrich, Scott Momaday, and Joseph Bruchac. These novel-
ists explore how Native people construct the environment and landscape
that contribute to the story and often become a character in the story
"between you and the earth. One look and you know that simply to
survive is a great triumph, that every possible resource is needed, every
possible ally-even the humblest insect or reptile" (Bounds and Swartz
1998: 78). The landscape plays a vital role in their work. Many environmental issues affect natives to the southwest, such as mining, ranching,
development, loss of native plants and animals, and water diversion to
other inhabited areas. Their novels explore the relationship between land
injustice and native people.

Leslie Marmon Silko's Vision of the Environment for the People and Land in *Ceremony*

Leslie Silko uses nature in *Ceremony* to define the landscapes of the characters and explain how those landscapes are symbolically tied to the hero's regeneration and communicate the essential core of human existence. This event contrasts the attitudes of Euro-Americans and Native Americans toward nature and reveals the natives' alienation from their environmental heritage. As Edith Swan has demonstrated the significance of symbolic geography about Tayo's regeneration in the "the spatial/temporal paradigm of the world that Spider Woman fabricated for the Laguna forms Laguna symbolic geography" (Swan 1988: 229). This symbolic geography guides Tayo, the novel's mixed-blood protagonist, on her spiritual journey to health and peace. It also has a crucially essential literal landscape, which Silko characterizes as a naturalist conspicuous feature of her surroundings.

In her depiction of Tayo's first interactions with the Navajo medicine man emphasizes the importance of a similarly historical relationship to the place where he lives:

> They keep us on the north side of the railroad tracks, next to the river and their dump, where none of them want to live. He laughed. They don't understand. We know these hills and we are comfortable here. There was something about the way the old man said the word comfortable. It had a different meaning-not the comfort of big houses or rich food or even clean streets, but the comfort of belonging with the land, and the peace of being with these hills. But the special meaning the old man had given to the English word was burned away by the glare of the mirrors and chrome of the wrecked cars in the dump below. (Silko 1986:117)

Despite the decay of his immediate environment and the homeless he sees every day, Betonie maintains a strong feeling of inscape and teaches Tayo about a sense of place. The protagonist in Silko's novel displays the worth of the land, which is why place or landscape in the narrative works this way. Silko's other account of nature covers land that his ancestors initially inhabited, and Tayo bemoans white culture's occupation of the area in the early 1900s. When government-seized tribal lands are sold to ranchers and loggers, the land is stripped for profit, and the animals are slaughtered for sport:

...the logger had come, and they stripped the canyons below the rim and cut great clearings on the plateau slopes. The logging companies hired full-time hunters who fed entire logging camps, taking ten or fifteen deer each week and fifty wild turkeys in one month. The loggers shot the bears and mountain lions for sport. And it was then the Laguna people understood that the land had been taken, because they couldn't stop these white people from coming to destroy the animals and the land. (Silko 1986: 186)

This passage is significant because it portrays Tayo's recognition of this cultural history, which leads to symbolic defiance of miners, bomb manufacturers, and destroyers. Miners, bomb builders, and destroyers are detrimental to the environment, which means they destroy culture and community and the potential of a bond in and with a place in his memory.

From this point forward, *Ceremony* reminds of some history that involves the confrontation with Tayo's drinking buddies, who turn out to be the agents of witches at the old uranium mine identified as the place of witchery. As Tayo's grandmother watched the dazzling brightness of the first atomic bomb detonation, this witchery jeopardized native people's survival and brought the earth to the edge of destruction. In the final part of the *Ceremony*, Tayo has a vision that:

He had been so close to it, caught up in it for so long that its simplicity stuck him deep inside his chest: Trinity Site, where they exploded the first atomic bomb, was only three hundred miles to the southwest at White Sand...there was no end to it; it knew no boundaries; and he had arrived at the point of convergence where the fate of all living things, and even the earth, had been laid. (Silko 1986: 245)

According to this passage, the atomic bomb, the ultimate expression of human and environmental damage, is blamed on witchery. Across all bounds, the atomic bomb kills all humans and living creatures. Tayo's freshly reconstructed Laguna Pueblo identity and community are placed within a modern uranium mining and nuclear technology system, and this vision is a crucial turning point in the novel.

As a result, Tayo realizes that he was correct in assessing the break-down of distances and time. He sees how native and non-native have become entangled in a web of the devastation of their communities and cultures, which affects their entire communities in the total sense of the word. As Tayo realizes, "from that time on, human beings were one clan again, united by the fate of destroyers planned for all of them, for all living things" (Silko 1986: 246). The scale of the witchery's final pattern of nuclear holocaust obliterates distinctions among races, nations, species, and geographical distances, bringing all earthly beings back into a worldwide circle of death.

Exploring the Interconnection Between the Land and Environment in Louise Erdrich's *Tracks*

Native people have strong ideas about nature and the environment, which are reflected in their writings. Like other North American Native authors, Louise Erdrich depicts the social concerns in her culture linked to the environment, causing native people to rise for an ecological cause. The novel *Tracks* by Louise Erdrich demonstrates how harming the environment became damaging to humans, with the lumber industry destroying flora and fauna and disrupting ecological equilibrium. The interconnectedness of nature and culture, notably the cultural artifacts of language and literature, is the subject of ecological equilibrium. As a critical stance, "it has one foot in literature and the other on land; as a theoretical discourse, it negotiates between the human and the nonhuman" (Glotfelty 1996: xix), and Louise Erdrich shows her characters entirely in contact with their physical environment. Nanapush, a character in Tracks, is a traditionalist who feels the loss of land keenly that "I heard the groan and crack, felt the ground tremble as each tree slammed earth. I weakened into an old man as one oak went down, another and another was lost, as a gap formed here, a clearing there, and plain daylight entered" (Erdrich 1988: 9). He knows instinctively that a loss of sovereignty results in environmental damage. The Allotment Act has depicted in action by Nanapush, representing the measuring and sale of land, the extraction of its resources, and the impact of these acts on the novel's characters. As the deforestation of the land surrounding the reservation begins to sell to the lumber cartels, trees in all their forms serve as symbols for native people throughout *Tracks*.

The extended family of Fleur Nanapush, Margaret's sons, Eli and Nector, have their access cut off in the form of surveyor lines and flags, plat lines, fences, telegraph poles, and other markers of the so-called progress imposed by the act. In the contentious territory of the post-allotment reservation, where the discourses of land, law, and natural environment collide, the land is seized and deforested. Nanapush describes "twisted stumps of trees and scrub, the small, new thriving grasses which had been previously shaded...the ugliness, the scraped and raw places, the scattered bits of wood and dust" (qtd. in Fitzgerald 2015: 60). The wheels of large lumber wagons crisscross the reservation roads, ripping the dirt wide. The novel establishes a link between Native history and their environment. Erdrich reacquaints characters with the Ojibwe past by referencing natural landmarks or climatic events to shed insight into the tribe's history. The Ojibwe recall heritage, living near the ocean, and the protagonists are fascinated with the Matchimanito Lake throughout the novel. In the narrative, the interaction and reliance of characters on the Lake and its environs are clear. As Pauline writes about Fleur and Eli, "In the morning, before they washed in Matchimanito, they smelled like animals, wild and heady, and sometimes in the dusk their fingers left tracks like snails, glistening and wet" (Erdrich 1988: 72).

Characters get identified with the setting. The reservation setting is consistently mentioned, emphasizing the importance of the real world. They were compelled to live close together on the reserve, and the winters brought "bitter punishment" in the form of the disease (Erdrich, 1988: 2). Because of the crowded environment, the disease spread swiftly, and the majority of them died. Erdrich also cautions about the potential for eco-catastrophe as a result of human acts that degrade nature. Ecocriticism also emphasizes the importance of preserving the natural-human balance. Snow, water, and a lake appear throughout the work as motifs that assist the characters in understanding how humans interact with the environment and how to survive in the event of an environmental calamity. The lake is a notoriously wild and dangerous location. Fleur is portrayed as a personification of nature, as she lives in the outdoors and is strongly connected to it. In the town, it is also said that Fleur killed Napoleon by drowning him in the lake. As Pauline describes Fleur, "her glossy braids were like the tails of animals, and swung against her when she moved, deliberately, slowly in her work, held in and half-tamed" (Erdrich 1988: 18).

The government's attempts to buy Indian land and then sell it to logging companies' intent on harvesting the trees continue to destroy tribal lands and traditions, as Nanapush rebels against the government's attempts and says:

I've seen too much go by-unturned grass below my feet, and overhead, the great white cranes flung south forever. I know this. Land is the only thing that lasts life to life. Money burns like tinder, flows off like water. And as for government promises, the wind is steadier. (Erdrich 1988: 33)

Tradition is connected with preservation in this context. A society that regards land primarily as a path to immediate riches threatens shared land, the inherited tribal past. The land has the power to pull them back to their roots, and it manifests itself in their blood "as if it runs through a vein of earth" (Erdrich 1988: 31). Pauline could never break this bond; after initially leaving for Argus, she comes home, recognizing the clues she receives from nature. Similarly, Margaret, the old lady and mother of Eli Kashpaw, "wanted a place right there that she could trust for her old age" (Erdrich 1988: 57). It is the repository of their lives, cultures, identities, and trust. Native societies revered nature and saw man as a servant of nature. The concept that the Earth is alive and that she is their protector mother establishes a special attachment with her. North American Native people develop a new relationship with the environment by recognizing these principles, realizing that our activities either destroy or save this world, essential for our survival.

Longing for Nature as Keeper of the Earth in N. Scott Momaday's *House Made of Dawn*

Native Americans are frequently depicted as environmentalists, keepers, and devotees of a Mother Earth goddess. These fabricated native environmental attitudes serve as a symbol for Indians and non-Indians to communicate a counter-narrative to the pervasive Western, techno-industrial attitudes toward and treatment of the land. Momaday represents the scenery of the southwest in *House of Made of Down*, an area of the country that is also important to the environment:

> There was a house made of dawn, it was made of pollen and rain, and the land was very old and everlasting. There were many colours on the hills, and the plain was bright with different-colored clays and sands. Red and blue and spotted horses grazed in the plain and there was a dark wilderness on the mountains beyond. The land was still and strong. It was beautiful all around. (Momaday 1999: 1)

The fundamental description of the environment and land relates the scene to native people because it is a condensed portion of a much lengthier song that native people chant. *House Made of Dawn* refers to the environment and earth and occurs "when the characters internalize images of the land by means of the symbolic acts of singing and story-telling" (Martin 2011: 62). This reminds Abel, a hero, of the land that is connected with the ceremony of "clearing the irrigation ditches in the spring. It is an imitation of water running through the channels, a magic bid for the vital supply of rain, and a ritual act to prevent the harvest from being influenced by evil powers" (Martin 2011: 63). On the other hand, nature exudes a sense of sadness and loneliness, foreshadowing the difficulties Able would face from the start.

Native People who are urbanized and assimilated into the white culture in North America use this term to describe those who are reluctant to change. Francisco is a longhair who follows historical customs and rites. Francisco's ways adapted to the environment and community he lives in because he does not have to live beyond the Jemez Pueblo. The community admires Francisco for his manner of living, and he is one of the environment's caretakers by imitating the conquerors' actions and gestures. Abel's Navajo acquaintance in Los Angeles, Ben Benally, feels connected to Abel because of their comparable cultures and backgrounds in the high desert region. Their relationship to the environment and land appears to comprehend that "similarities in culture derive in turn from similarities in the landscapes out of which the cultures emerged" (Martin 2011: 63). Benally compares his reservation upbringing with a description of life city environment:

> You have to watch where you are going. There's always a big crowd of people down there, especially after it rains, and a lot of noise. You hear the cars on the wet streets, starting and stopping. You hear a lot of whistles and horns, and there's a lot of loud music all around. Those old men who stand around on the corners and sell papers, they are yelling at you, but you cannot understand them. I cannot, anyway. (Momaday 1999: 140)

Benally, like Abel, was reared on a reserve and retained ties to his tribe's land and traditions. It's a place where native people assess their environment and decide where they belong. Integration failed with a person so closely bound to the environment of his birthplace. This man needs to live according to traditional conservative rules to define himself and give his life meaning. The federal programs intended to abolish reservations and integrate the people into mainstream society and economy by providing transition benefits such as job training and health care.

Abel has reached a stage of complete breakdown and devastation, a threshold of near-self-destruction that appears to be required to initiate a healing and environment-sighting process. His deafening blindness and lack of sense of direction are further demonstrated by the following:

> You felt good out there, like everything was all right and still and cool inside you, and that black horse loping along with the wind...you were coming home like a man, on a black and beautiful horse...and at first light you went out and knew where you were. And it was the same, the way you remembered it, the way you knew it had to be; nothing had changed. (Momaday 1999: 154)

Benally's vision of Abel's vision is a moment of eternal unity and being one with the land that leads him to heal in the context of the countryside. Able reconciles his individual and tribe identities, and the story is intended to be a verbal affirmation of the environment and tribal spirit.

APPROACHES TO LAND, NATURE, AND WILDERNESS IN JOSEPH BRUCHAC'S *LONG RIVER*

These issues look into uranium mining and its effects on people and the environment. The destruction of the natural world by dams, pollution, oil wells, and uranium mines has been a genuine threat to native self-determination throughout the century. Along with their counterparts in uranium-rich native regions, these people were victims of uranium mining. The tribe deals with a dam that shuts off the river and symbolizes the elimination of traditional religion and the government and corrupt corporations seeking to profit from Indian territory and resources. Younger Hunter, a character, has proceeded on his perilous trip to confront the terrifying Ancient Ones and prevent the massacre from destroying his reserve land in *Long River*. The Abenaki people emphasize the interconnection of all things, emphasizing the importance of developing a reciprocal relationship with the natural world. As Jace Weaver describes, North American Native people are removed from their reservation and deprived of numerous landscapes "that are central to their truth and identity, lands populated by their relations, ancestors, animals and beings both physical and mythological" (Madsen 2015: 12).

Animal people, according to a character such as Bear Talker, are just like us:

> Carry life and are made of the earth. Those bodies go back into the earth. Then we have the breath spirit that enables our bodies to move. That comes from Ktsi Nwaska and is carried by the wind. It goes back into the wind when our bodies return to the earth. When we hunt an animal, we do not harm that part of it. That returns to the owner creator...there is more than...no bear is exactly like any other bear. No human is exactly like any other human. When the body dies and the breath leaves that spirit steps free. It does not die but goes to the Owner Creator. (Bruchac 1995: 25)

These ties reveal Native people's tribal nature. Images from their traditional culture teach them about the relationship between people and animals, Native and non-Native, and the rest of the planet. Bruchac enlivens their imaginations and strengthens the circle of the river that binds Natives to the landscape and culture through his story of remembering his homeland and environment. This is the representation of the Abenaki world's harmony and balance through storytelling. As Joseph Bruchac describes that:

> Native stories play in reminding people that all life is a circle, dynamic and moving in a continuous cycle in contrast to the European view that life is linear, and that the goal is progress. In his retelling of the Abenaki story of Gluskabe and the Wind Eagle, Bruchac explains how a simple story illustrates a severe environmental problem caused by human shortsightedness and shows the necessity of seeking wise counsel and a remedy to make things right again. It is this process, which the stories teach us, a process we must follow as individuals, as leaders, as nations... Feeling of guilt does nothing. Only action to restore the balance is the proper response. (Ricker 1996: 160)

In this setting, Bruchac observes how Young Hunter and other characters preserve their culture and environment by listening to the native elders around them, implying that humans cherish the soil, wildlife, and animals the same way as ancestral Native people did. By getting close with a hunger for the killing of Native people, Young Hunter detects the presence of two demonic entities in the forest, one the size of a hill and the final survivor of the Ancient Ones. Young Hunter, Willow Woman,

Ancient Ones, and others discovered a means to blend the stories of their cultural representation with the realities of their experiences after becoming estranged from the social and marginal places around them. In this context, cultural conservation between human, animal, nature, and mythical character of a storied is a core point of this novel. Bruchac specifies its conversation to deal with the monsters where Medicine Plant, Bear Talker, Willow Woman, Dog, and Young Hunter defeat the monster on the top of the mountain in undergrounding physical spaces and water sources to search the right path:

> They know you have been warning the people. They know you are very brave. So, they have sent me to find you so that you can help us to defeat the monster. You can trust our deep-seeing people: Medicine Plant and Bear Talker and Young Hunter...my people know that a great monster, one that hunts the human beings, is coming toward them. (Bruchac 1995: 262)

Young Hunter, along with other Native people and communities on the reservation mountain, faces foreboding dangers and fights to maintain the ecosystem and their representation both on and off the reservation after returning to their tribal world into balance. In *Long River*, the character, Younger Hunter has continued on his harrowing journey to face the terrible Ancient Ones and prevented the destruction of the massacre coming to his Abenaki people. The Abenaki people underscore the interconnectedness of the things that highlight the value of cultivating a reciprocal relationship with the natural world. As Young Hunter realizes the sources of the real power of their reservation land are the body of the bone-spirit "that can leave the body and return or continue to the sky land, and then there is the fourth one. And that one may return in the shape of a new one may some of the memories from that earlier life" (Bruchac 1995: 30). Bruchac draws the power of internal connection to defeat white culture in the urban area and sees their peace to return to the reservation land where Abenaki people live in harmony with nature and life-giving river, which restore native people to represent their culture and locate the bones of Native elders by conducting ceremonies on and off reservations. This offers two crucial interventions: they highlight a regionally intense and internationally vast place-connection and emphasize the need for historicist methods to Native literature studies that can address environmental applications of intellectual practice. These interconnections uncover Native people are part of the tribal nature and the images of their traditional culture teach them about the relationship between people and animals, Native and Non-Native and the rest of the earth. *Long River* uses the traditional and contemporary experience to explore urban identity as a link to the practices, beliefs, and culture of their representation.

CONCLUSION

The question of climate change and environment are closely linked to the indigenous people of North America. This is clearly visible in the analysis of the texts taken in this article. What is more crucial is its engagement with larger issues on environment and ecology visible in the Asia–Pacific region. Environmental concerns along with rapid industrial growth have led to destabilizing effect on the inhabitants of this region. The construction of dams, nuclear stations, chemical factories, building of roads, Uranium mines, railways on the areas closer to the reservations or tribal lands has not only impacted the indigenous population but also accounted for rapid deforestation and environmental degradation. Policies are being implemented across nations to minimize the same and Asia–Pacific has a major role to play. The texts discussed in this paper are relevant in understanding the global cause on environment and climate change. They posit not only a nativist position but also open the space for integrating a comprehensive approach to make a connection between the land and its inhabitants.

REFERENCES

Bounds, Susane and Patti Capel Swartz. (1998) 'Women's Resistance in the Desert West' in Murphy P. D. (ed.) *Literature of Nature: An International Sourcebook.* Chicago and London: Fitzroy Dearborn Publishers, pp. 77–83.
Bruchac, Joseph. (1995) *Long River.* Colorado: Fulcrum Publishing.
Carpenter, Bird Delores. (1994) *Early Encounters: Native Americans and Europeans in New England from the Papers of W. Sears Nickerson.* Michigan: Michigan State University Press.
Chalfant, William. (2002) *Cheyennes and Horse Soldiers: The 1857 Expedition and the Battle of Solomon's Fork.* Norman: University of Oklahoma Press.
Deloria, Vine. (1992) 'Trouble in High Places Erosion of American Indian Rights to Religious Freedom in the United States' in Annette J. (ed.) *The State of Native America: Genocide, Colonization, and Resistance.* Boston: South End Press, pp. 267–290.

Erdrich, Louise. (1988) *Tracks*. New York: Henry Holt.

Fitzgerald, Stephanie J. (2015) *Native Women and Land: Narratives of Dispossession and Resurgence*. Albuquerque: University of New Mexico Press.

Glotfelty, Cheryll. (1996) 'Introduction: Literary Studies in an Age of Environmental Crisis' in Cheryll G. and Harold B. (ed.) *The Ecocriticism Reader: Landmarks in Literary Ecology*. Athens and London: The University of Georgia Press.

Lewis, David Rich. (1995) 'Native Americans and the Environment: A Survey of Twentieth-Century Issues', *American Indian Quarterly*, 19 (3), pp. 423–450.

Madsen, Deborah L. (2015). *The Routledge Companion to Native American Literature*. New York: Routledge Publication.

Marder, William. (2005) *Indians in the Americas: The Untold Story*. San Diego: Book Tree.

Martin, Holly E. (2011) *Writing Between Cultures: A Study of Hybrid Narratives in Ethnic Literature of the United States*. Jefferson, NC and London: McFarland & Company, Inc. Publishers.

Momaday, N. Scott. (1999) *House Made of Dawn*. New York: Harper Perennial Modern Classics.

Porter, Joy. (2012) *Land and Spirit in Native America*. Santa Barbara, CA: Praeger/ABC-CLIO.

Ricker, Meredith. (1996) 'A MELUS Interview: Joseph Bruchac', *MELUS*, 21 (3), pp. 159–178.

Robyn, Linda. (2002) 'Indigenous Knowledge and Technology: Creating Environmental Justice in the Twenty-First Century', *American Indian Quarterly*, 26 (2), pp. 198–220.

Santoro, Nicholas. (2009) *Atlas of the Indian Tribes and the Clash of Cultures*. New York: iUniverse, Inc.

Schweninger, Lee. (1993) 'Writing Nature: Silko and Native Americans as Nature Writers', *Melus*, 18 (2), pp. 47–60.

Silko, Leslie Marmon. (1986) *Ceremony*. New York: Penguin.

Swan, Edith. (1988) 'Laguna Symbolic Geography and Silko's *Ceremony*', *American Indian Quarterly*, 12 (3), pp. 229–249.

Future Impacts of Climate Change on the Lives and Livelihoods of Indo-Fijians

Kate Martin

"Global forces have often had detrimental impacts on the Pacific islands. Colonization, island-scale phosphate mining, nuclear weapons testing, the Second World War, the Cold War, globalization and trade liberalization have all wrought significant political, economic and cultural changes in the region. Yet… the most dangerous to the Pacific islands is climate change" (Barnett 2005). As CO_2 concentrations in the atmosphere have increased by 30% since 1750 a mixture of farming, overfishing, industrialization, land clearing, and deforestation have become major sources of greenhouse gas emissions. The Pacific Islands have resultantly begun to exhibit an array of environmental problems that increase their vulnerability to the effects of climate change, the list of which is extensive: "land degradation, such as soil nutrient depletion and soil loss; deforestation due to logging for timber exports, clearing for agriculture and fuel-wood collection; biodiversity losses across a range of terrestrial and marine flora and fauna; depletion of freshwater resources as a result of saline incursions and contamination from urban, agricultural and industrial sources; and

K. Martin (✉)
University of Central Lancashire, Preston, UK
e-mail: KMartin5@uclan.ac.uk

© The Author(s), under exclusive license to Springer Nature
Switzerland AG 2022
N. J. P. Alsford (ed.), *Pacific Voices and Climate Change*,
https://doi.org/10.1007/978-3-030-98460-1_9

191

coastal and marine degradation, including coastal erosion, coral loss and coral bleaching, contracting artisanal fisheries and pollution of lagoons" (Barnett 2005). The scale of the ongoing destruction can easily be linked to the changing human presence in the region and beyond (Devadason et al. 2019) but is also due to the increasingly locust-like behaviour of mankind, as we thoughtlessly abuse the natural world, either ignoring or failing to recognize the increasingly devastating impact it has on our brothers and sisters thousands of miles away.

For the indigenous and minority, ethnic people of Oceania climate change presents a very real danger to the livelihoods and cultures that have been established and evolved over generations. However, for those who live beyond the Pacific Islands the image of ethnic homogeneity that is often presented in the media is almost entirely inaccurate and rarely represents the true ethnic and cultural makeup of the region. As migration, colonization, war and now climate change shape and mould Oceania we can no longer cling to the distant Anglo-American orientalism images of "tropical islands, bare breasted women and primitive culture" (Batman 2015), instead we must wake up to the reality of nations who have been left at an economic disadvantage due to decades of colonial oppression and the leaching of natural resources by Western nations (Oliver-Smith 1996). As king tides rise and crash upon the shores of small island nations across the South Pacific and the costs of climate induced natural disasters increase yearly for those countries struggling to develop their economies (Acting on Climate Change & Disaster Risk for the Pacific 2013), the outlook is bleak. But what of those non-indigenous ethnic groups who settled within the region generations ago, who have often struggled to achieve equality and build lives in distant lands where they share no linguistic, cultural, or historical links with the indigenous population. Without TEK (Traditional Ecological Knowledge) (Bryant-Tokalau 2018) how are they to cope with the complex challenges climate change will present them in the decades ahead? That is one of the questions this study asks about Indo-Fijians.

As a diaspora group, Indo-Fijians are one of today's most significant examples of a bygone colonial era where indentured labour of Indians occurred en masse in every corner of the British empire. Whilst the forty years of indentured labour (1879–1916(20) has passed from living memory, those descendants of girmityas (indentured labourers) retain the culture and language of their forebears and in doing so have produced a new identity (D'Souza 2001). However, finding a foothold in Fijian society has been something of an uphill battle, where inequality has shadowed their every step and basic rights have, like their many diasporic cousins, been drip fed to them after considerable social pressure. During the colonial period the rights of indigenous Fijians garnered more focus from the colonists who gradually implemented their "political, cultural and socio-economic supremacy" (Kumar 2012), which can still, to some extent, be observed to this day. Such policies for the protection of indigenous Fijians were of course important for preserving the traditional systems and power structures of Fijian society. Yet, it is evident that few considered how importing labour from India for the purposes of economic development would change the social dynamics in Fiji, especially considering how until the 1960s Indo-Fijians were in the majority, outnumbering indigenous Fijians. From the late 1960s to 2000 Indo-Fijians found themselves in a one step forward, two steps back system. As Fiji gained independence in 1970 and began to experience the trials and pitfalls of a modern democracy, a new generation of Indo-Fijians experienced a more democratic way of life. 1977 marked the first time the Indian-led opposition won a majority in the legislature. In what should have been a win for equality an unfortunate reality set in and the opposition party determined that it would not be possible to form a government in fear of the indigenous Fijians' reaction to an Indian leader. It would be another 10 years (1987) before an Indian would become Prime minister, spending only one month in office before being removed by coup and imprisoned. This stoked the fires of discontent on both sides of the political coin, causing widespread racial violence, as well as protests and demands for the Prime minister's immediate return. Over the following six months a second coup and a new constitution rocked the land and placed indigenous Fijians in an unequal power balance with Indo-Fijians. Across the ensuing years and months Fiji became more and more inhospitable to Indo-Fijians as they were subject to threats on their unions, firebombs aimed at their businesses/places of worship, arrests and assaults on members of the community, and few opportunities to take part in the governance of their country. Students, farmers, and other members of the Indo-Fijian populace protested and attempted to utilize every political channel to make change. In 1997, a new constitution was adopted finally giving way to a new set of elections and another Indo-Fijian Prime Minister whose term in office would be little longer than his predecessor. By failing to recognize the precarious social position Indo-Fijians have been left in, which has often led to persecution in periods of political upheaval, one cannot help wonder, if, when climate change begins to put pressure on such a fragile system, which continues to lack social cohesion, whether Indo-Fijians will be first to fall through the cracks of society.

During 2000, a political coup and riots incited by misguided ethnic Fijian nationalism caused the Indo-Fijian Prime Minister, Mahendra Chaudhry to be ousted within a year of taking office. Such action demonstrated how it was not only a lack of opportunities that was pushing Indo-Fijians to emigrate to Australia, New Zealand, Canada, the United Kingdom and the United States, but also the apparent institutional racism, which left them at a marked disadvantage and would hinder their ability to prosper (Lal et al. 2008). If these conditions were to be sustained then the consequences of climate change, such as reduced resources and access to land could have severe repercussions and result in larger scale displacement and migration. Despite constitutional changes designed to ensure equal rights, the exodus of Indo-Fijians has continued (Indo-Fijians - Minority Rights Group 2017), which can only be interpreted as either a lack of faith in the system caused by the embedded historical hostility shown towards Indo-Fijians or the continued reduction of employment opportunities. Failure to meet the disaffected populations expectations of twenty-first-century life will inevitably perpetuate the situation.

Whilst the links between minority ethnic groups and climate change in the Pacific Islands are still in their formative stages, there are well-documented examples of the disproportionate impacts of climate induced crises in other regions such as the case of Hurricane Katrina, which due to racial inequality resulted in the unequal distribution of resources between the majority Caucasian community and the African American residents in the aftermath of the disaster (Baird 2008). If similar circumstances were to befall Indo-Fijian communities, particularly in coastal areas where access to land has always been a source of tension, this could lead to internal displacement as those indigenous rights of access to traditional lands are protected, and if tenancies for Indo-Fijians fail to be renewed, the Fijian government will have to find new ways to help Indo-Fijians to cope. The International Organisation for Migration has speculated that under Prime Minister Bainimarama, Fiji has seen a number of land reforms, which have led to Indo-Fijians returning to the islands, but questions have arisen about how secure and successful such changes will be, if, in the future, their access to land is threatened or perceived to be insecure, in which case, large-scale migration will again be utilized as a method of climate change resilience (Effects of Climate Change on Human Mobility in the Pacific and Possible Impact on Canada 2016). Considering the number of relocations already expected in coastal regions due to the environmental hazards caused by climate change we will probably begin to see the first examples of climate induced internal migration within the next decade. This will however only go so far as "there are currently no clear plans for the relocation of Fijian communities of Indian descent. Government relocation support is reserved for iTaukei communities despite the fact that these groups are equally vulnerable to the consequences of climate change" (Neef 2019). For the majority of today's Indo-Fijians who are concentrated in the sugar belt or in the coastal cities and towns of Viti Levu and Vanua Levu few have inland connections outside of the aforementioned areas, as for generations, they have built their lives and livelihoods in more densely populated areas, with little access to the traditional indigenous knowledge and understanding of their environment. This issue hindered the transferal of TEK, which can only be accumulated over time and exposure to surviving off the land. As we will mention later in the chapter, Indo-Fijians are by and large a well-educated group who place great importance on education resulting in many holding advanced degrees or having attended college. A high percentage of Indo-Fijians are employed within traditional agriculture, fishing, clerical, retail, transportation, crafts, business, finance, and politics occupations (Indo-Fijians - Introduction, Location, Language, Folklore, Religion, Major holidays, Rites of passage 2004). Unfortunately, these professions are either directly affected by climate change or will need to provide the resources and policy changes necessary to mitigate its impacts. Experiencing environmental or climactic events will be increasingly traumatic but, may also present important opportunities to improve the Indo-Fijian population's understanding of TEK and could result in climate resilience innovations. Identifying methods and means of resilience will rely on the sharing of information by indigenous Fijians with a willingness required to move forward together and not allowing old, non-productive ethnic politics, and segregation to stymie progress.

To further understand the unique circumstances of the Indo-Fijian people and how this will inform their future we must delve into the related literature. Whilst it is difficult to find information that directly links the ensuing climate crisis with the inequalities felt by ethnic minorities in the Pacific Islands there is reference material available concerning the lives and livelihoods of Indo-Fijians and the various upheavals and crises they have experienced, particularly over the last fifty years. This paper, therefore, is something of a jigsaw puzzle as we piece together the systemic inequalities of Fiji's political, economic, and social systems with the outcomes of previous climate induced disasters. We will also draw comparisons with similar circumstances that have befallen ethnic minorities in areas such as the United States, and also question whether, as the climate changes, Indo-Fijians will suffer disproportionately to their indigenous counterparts that is comparable to other ethnic minorities in areas at risk of large-scale climate-induced disasters. The belief of this paper is that such inequalities do exist and that without a framework being put into place by both indigenous and Indo-Fijians to mitigate the impacts of climate change many could find themselves in a similar position as that which was experienced during the coups and political unrest of the 1980s and 2000, with few ways out of the situation other than migration.

Lives & Livelihoods

Farming

"Agriculture is a key pillar of the Fijian economy: 65% of Fijians derive at least part of their income from agriculture and the agriculture sector employs 45% of the population" (Sleet 2019). Indo-Fijians have contributed to the agricultural economy of Fiji not only through years of indentured service on sugar plantations and in their tenure of the land thereafter but also through their carrying new and diverse crops with them across the sea. If you were to walk amongst the market stalls of Suva, Nadi, or Tavua you would find vendors selling spices, rice, lentils, beans, chickpeas, karala, chillies, tomatoes, cucumbers, eggplants, okra, mangoes, and pumpkin, all ingredients found more commonly in traditional Indian cuisine (Sundaresan 1985). Though today both Indian and indigenous Fijian farmers grow these products amongst an assortment of others they have inspired a change in food consumption across the larger islands, which is now being threatened by climate change as these crops for the most part lack both salt water and drought resilience.

Scientists have stated for decades that there is a link between warmer ocean temperatures in the South Pacific and the increased frequency and intensity of cyclones, storms, and flooding. Presently, around 69% of external hydrological shocks (economic and environmental extremes) in the Pacific are attributable to storm surges, landslides, cyclones, and flooding which are gradually increasing in extremity (Devadason et al. 2019). Cyclone Winston confirmed this by destroying huge tracts of land including everything that had been planted within it. As temperatures and rainfall levels change, commercial sugar farmers (an area of agriculture still dominated by Indo-Fijians) in particular, are feeling the strain (CCTV 2013). Coastal inundation has been recognized as one of the most prevalent and dangerous threats to Pacific island shorelines and is one which will continue to take inches off the land in the years to come (BBC 2020). "Agricultural activity will be affected correspondingly - not only will sea-level rise decrease the land area available for farming, but episodic inundation will increase the salinity of groundwater resources" (Keener et al. 2013).

As yields reduce due to a mixture of extreme weather events and climate induced crises, farmers have had to diversify and adopt climate-resilient techniques. With such a large proportion of the Indo-Fijian and indigenous Fijian population lacking the awareness of ancient Fijian farming methods, which could be more successful, the United Nations Development Programme have stepped in to coordinate workshops introducing a new "triple layered integrated farming system with some short-term crops of three to four-month maturity, medium-term crops such as taro and cassava, and long-term crops" (Narang 2018). This has gone some way to securing a crop rotation that can be resilient to disasters, but it is not able to deal with the conditions of rising sea levels.

By ensuring that Indo-Fijians are included in the application of traditional indigenous knowledge and methods of climate resilience (though their livelihoods will still be affected by climate change), we can ensure that they are not disproportionately impacted in comparison to indigenous Fijians. However, something that we must acknowledge is those methods of resilience to extreme conditions passed down through the generations that trace back to their roots in mainland India. One of the most obvious being back-up subsistence gardens which can be used not only to subsidize the food brought into the home, but also to ensure a supply during disasters (Leonard 2019). By continually growing fresh produce and utilizing traditional food preservation methods such as pickling and making chutneys, a practice handed down to modern-day Indo-Fijians, they can ensure the availability of food during times of drought or flooding (Balachander 2012). This is a behaviour which can be observed in both the rural Indo-Fijian communities as well as those living in more urban areas. Weaving together methods for resilience from across the sociocultural spectrum will be invaluable to ensuring the livelihoods of Indo-Fijians going forward.

One of the most significant concerns for farmers will be the salinization of groundwater, which will not only destroy one or two yields in a year but make land untenable, thereby destroying any chance of achieving commercial agriculture and eventually forcing farmers from the land and into towns and cities (Keener et al. 2013). As resources reduce and reliance on imports increases, one can speculate that those Indo-Fijians who do not own the land they work on will not have their tenancies renewed (Tegunimataka and Palacio 2021) to ensure that indigenous Fijians can directly access the land and agricultural economy, which would, at this stage, be entirely detrimental to both parties who are now co-dependent, even if they do not (yet) realize it. We can already observe the results of land leases expiring without renewal as large numbers of rural migrants begin to flock to Fiji's urban areas, which are growing faster than the rate of population. One of the most significant waves of migrants took place in 2001 when 13,100 leases expired, affecting around 22,000 people (The Pacific Way Story - Struggling for a Better Life, Squatters in Fiji 2014). As a result, numerous informal settlements are springing up in and around Suva and Port Moresby, with such residential areas now housing more than half of the urban population (Firth 2006). The new Fijian favelas will have a significant impact on the environment as the lack of amenities will lead to human waste of all kinds failing to be disposed of in an appropriate manner and instead being left along the coast line leading to the death of coral reefs and the harm of various flora and fauna, whilst further inland it could lead to the contamination of freshwater lenses or lagoons (The Pacific Way Story - Struggling for a Better Life, Squatters in Fiji 2014). This inadvertent destruction of the natural world will have ramifications for Indo-Fijians and indigenous Fijians alike. Forcing people who are born farmers and have the knowledge to help Fiji's agricultural economy survive even in times of crisis to scratch out a living on the edges of towns and cities is a waste of a valuable resource and will inevitably result in a decline in the numbers of people able to productively work the increasingly fragile land.

Fishing

When one conjures up an image of fishing in the South Pacific you may imagine the spectacular colours of the coral reefs, teaming with life in a perfectly balanced ecosystem. Or you may picture the warm crystal-clear waters glistening in the midday sun, almost untouched by humanity, and entirely apart from the modern world. A place where woman, man, and creature live together in harmony, but sadly such a vision is a fallacy. As the ocean temperatures rise coral reefs are calcifying, fish are dying, and the Pacific Ocean is beginning to look more like a Tim Burton film than a Disney utopia, but, like all good Tim Burton movies, there are signs of hope beyond the darkness. One must hope that this will be the case for those Indo-Fijian fishing communities who have coped with the challenge of respecting indigenous fishing culture and must now face the threat of climate change's impact on their livelihoods. In a recent study of Ba province (Fiji) by Chinnamma Reddy, the shift away from farming sugar cane into fishing was noted as shifting the socioeconomic and socio-cultural base of the region (Reddy 2019). It is now possible to recognize five groups of Indo-Fijians working in the fishing industry; "1) Machuas who are the actual fishermen; 2) Gayas who own the fishing resources and usually have two or three people accompanying them on fishing trips; 3) Dayas are usually someone who has a fish retail business and provides the necessary resources for fishing and would occasionally or never go on trips; 4) Maya, as usually the middlemen, buying directly from the Machua's and the Gaya's to resell for a profit, and 5) Jayas' who normally purchase from the Maya's, re-bundle the catch and sell them at a higher price often to other retailers, restaurants or hotels" (Singh 2008).

As so many people are now reliant upon ocean produce to sustain themselves and their families, the increasing possibility of the ocean ecosystem crashing is likely to have devastating consequences for Indo-Fijians. The fish population around the coral coasts of Fiji are already decreasing due to overfishing. This has contributed to the excessive growth of sargassum and other filamentous algal species, which has over-whelmed certain corals in the reef system. After coral bloom events, local scientists were brought in to study the area, decisions were made to slow the decay of Fiji's coast. By establishing locally managed marine protected areas under community control and enforced tabu (bans) alongside other development strategies, the coral reefs of Fiji have begun to feel less pres-sure from direct human impact (Morrison et al. 2013); however, climate change will continue to bleach coral reefs and make the sea uninhabit-able for the creatures within if the wider global community continues to contribute to pollution and the warming of our planet.

The use of indigenous knowledge has been invaluable for sustaining Fijis reefs, but one cannot help wonder how regulating fishing is conducted and by whom, as this could lead to an inequality in fishing access for Indo-Fijians who continue to respect the customary fishing rights known as i-qoliqoli, in which indigenous Fijians are the owners and Indo-Fijians are the users. In dividing the two by characteristics of ethnicity, who resultantly has access to resources and ownership rights fails to be determined along equal lines, with participation in Fiji's fishing industry dominated by indigenous Fijians. The exclusion principles, which were previously utilized for the purposes of long-term sustainability, can also be used to shut out non-indigenous people in times of resource crisis (Cox et al. 2010).

More often than not the ability to access resources creates an environment of inequity, social exclusions, and conflict (Hollup 2000; Sherif 2013), which in a region already tainted with a history of ethnic tensions leading to coups, protests, and occasional violence can only be a source of angst for those currently residing and working in Fiji's coastal fishing settlements. In June 2021, a survey conducted across five fishing communities to discover how climate related events were impacting Indo-Fijian communities found that 59% of respondents were negatively affected by Cyclone Harold and 18% by bad weather. The cyclone damaged 26.2% of fishing boats and as insurance companies had refused to insure the traditional wooden boats which are used by Indo-Fijians, this combined with flooding reduced overall catches. The community is evidently struggling to cope with climate change and the obvious ethnic tensions are reducing mitigation capabilities (Manghubai et al. 2021).

If we were to look deeper at the gendered impacts for Indo-Fijian women who principally work as "go-betweens" by selling fish on the commercial market, it is clear that they will inevitably be the first to lose their income and economic autonomy, if, during times of crisis, there is a shift from commercial fishing to subsistence fishing. As changes in weather patterns are already concerning these women (Reddy 2018), it is of significant importance that Fiji finds more equal ways to distribute fishing rights to ensure Indo-Fijian fishermen and industry workers do not lose their livelihoods in the future, whilst ensuring that whichever choices they do make are sustainable.

Business

Indians are ubiquitous, with thriving communities spanning across the Americas, Europe, Africa, Asia, and Australasia. Part of their success is due to their adaptability, business acumen and strong sense of identity (Daye, 2009; Kumar and Steencamp 2013) whilst the remaining part can be linked to the somewhat unique place they held in the old colonial social order, wherein Indians were brought to colonial outposts as labourers but often found their place somewhere between the colonists and the indigenous population. Occasionally, this would come with more liberties, thereby socially isolating them from the experiences of the indigenous population, whilst they remained apart from their colonial overlords. However, Fiji is a notable exception to this as the indigenous Fijians were given access to the power structures and their traditions were to some extent maintained, whilst Indians were barred from participating in politics, owning land or accessing natural resources (Lal 2012). They were, until independence, considered to be third class citizens and after the departure of the British colonists they then became second class citizens who were continually denied equal rights under the law.

It quickly became evident that there were few professions left open to those Indo-Fijians who no longer wished to work on sugar plantations. Meaning once again that Indians would find themselves in the position of middle-man by setting up numerous businesses and enterprises. Initially Indo-Fijians supplied their own communities but as businesses have grown they have come to dominate the landscape of towns and cities. In a 2004 survey of 300 businesses based in Suva, Nadi, Lautoka, and Ba just over 80% were owned by Indo-Fijians who on average have a higher income than their indigenous counterparts and can make up to 428.5% more in certain cases (Reddy 2007). This is indisputable evidence of the industriousness and entrepreneurship found in Indo-Fijian communities. Nonetheless, it also highlights that to be one's own boss was the only way of escaping employment discrimination, meaning to this day that the majority of the small business sector, in particular family-run businesses, are dominated by Indo-Fijians. It is also evident in tourist resorts and hotels where back-office clerical positions in accounting, administration, record keeping as well as trade jobs are allocated to Indo-Fijians. These positions are usually well paid and require a higher level of education, which is often found amongst the Indo-Fijian populace who on average stay in school longer (Reddy 2007).

The position of Indo-Fijians as workers in the commercial economy is, for the most part, reliant upon having access to resources which can later be redistributed. As 80% of Fiji's land is indigenous owned (Kurer 2001), maintaining an efficient working relationship with landowners is of paramount importance in order to maintain the supply chain during times of crisis. Unfortunately, the vulnerabilities of small businesses such as restaurants, supermarkets, and general retail become apparent not only in times of economic crisis but also when ethno-nationalism comes to a head as was the case in the coup of 2000. As commercial and private properties were looted, smashed up and in rare occasions set alight (Trnka 2005), in Suva alone around 167 businesses were ransacked, the majority of which were owned by Indo-Fijians and Gujaratis (Trnka 2008).

If such violence can be incited for political purposes one can only begin to imagine the potential sequence of events which could take place if climate change continues on its current path. The extensive flooding, frequent cyclones and higher temperatures will lead to a compounding of issues resulting in damages to businesses and homes. They may also experience low agricultural yields which will drive up food prices in urban and rural areas. We could see major water shortages as groundwater is salinized and heavy rainfall becomes less frequent, this will inevitably lead to social tensions. As Indo-Fijian businesses have a higher financial capacity to weather a storm and rebuild in the aftermath, those feelings of inequality in affording enough food to feed the family and having access to the commercial markets in particular those of imports during natural disasters may lead to further coups, violence, and possible political scape-goating (Daye 2009) to take the focus away from climate change and onto a group who are simply trying to build a life, whilst facing the same ecological challenges as indigenous Fijians.

We know from the coups and general political instability that Indo-Fijians will choose to migrate, if they feel unprotected by the state and that their businesses, having been destroyed, are not worth rebuilding. Large numbers have chosen to use their savings to instead emigrate with their families. This causes significant "brain-drain" as well as damaging the commercial economy of the nation.

Migration

So, we know that Indo-Fijians will inevitably feel and potentially suffer the reverberations of climate change but what actions they choose to take in the wake of various climate crises is still to be determined. We can only suggest options which are perceived to be economically viable as well as socially acceptable. By re-tracing the movements of Indo-Fijians during previous crises and understanding their migratory patterns we can begin to observe the split between those who choose to leave and those who stay behind. Both decisions are multifaceted and are driven by a need for stability and safety. Something which, for Indo-Fijians, has not been assured historically and is still not ensured for the future. For Fiji 2001 was a year of turbulence and may have set off an unexpected chain of events which will seriously impact the future of the nation. Through political violence, further restricting access to land and resources and large-scale rural to urban migration, those Indo-Fijians with the capability to emigrate did so in waves, "between 1978 and 1986, just over 20,700 Fijians left the country at an annual rate of 2,300, in the ensuing decade (1987–1996) the rate more than doubled to 5,005 every year. The overwhelming majority, roughly 90%, of these departing citizens were Indo-Fijian" (Minority Rights Group 2017). Today there are around 313,798 (37.5%) (Refugees, 2017) Indo-Fijians, if the rate of migration continues at around 4,500 a year, then in just under seventy years there will be no Indo-Fijians left in Fiji. Considering how the rate of migration has not slowed and is speeding up, save the last two years due to Covid-19, we can speculate that if Fiji experiences consecutive years of extreme weather events causing further damage to their lives and livelihoods, those with the capacity to migrate will begin to do so in the belief that life in Fiji is no longer tenable due to climate change. If, akin to a tsunami, instability is sensed as the tides go out, few will wait for the waves to come crashing down upon them and will instead move away from the area if they are to avoid the damages and trauma to come. We must also monitor the barometer of those Indo-Fijians who have returned, but with an awareness that they will not stay if their access to land or other resources is at all affected by instability, this may also determine whether they will convince others to move on as well. Climate change may not have been the area of instability they were principally concerned about upon return to their chosen homeland, but it will inevitably lead to increasing the political frictions and ethnic tensions that remain just under the surface of Fijian society.

Migration being used as a method of resilience to changes in climate or the general environment is well documented throughout the Pacific region (Campbell 2014). For Indo-Fijians or for Indian diaspora in general, migration beyond the borders of their modern-day homeland was seldom done to escape or avoid changes in climate so there is little precedent for such migration patterns. Therefore, we have no choice but to focus on the socioeconomic factors which contribute to such choices. The loss of livelihoods, homes, and any sense of stability through climate induced damages, paired with prior movement of relatives due to political instability has led to the Indo-Fijian populaces' sense of roots being gradually gnawed away. The ensuing identity crisis and reduced sense of allegiance to a homeland that perceives them as lesser beings will inevitably result in large-scale migration that the Fijian government should be worried about. Yet, if migration is inevitable why should the global community worry about it and why not let the chips fall where they may? Why should we take note of what is happening thousands of miles away to a small group of people that we may never meet? The answer to all these questions and more is that we, as a global community must start to make changes now. With no template to work from we must attempt to understand the lives of earlier migrants both indigenous and non-indigenous, as well as observing how governments are moving people within and outside of their national borders to limit the effects of climate change on the more vulnerable citizens of their country. Finding suitable methods of resilience that do not rely on simply moving people away from a problem, such as sea-level rise, will reduce the anticipated trauma and avoid the devastating consequences that result from sudden large-scale human displacement situations such as experienced within Syria and Afghanistan. Although the associated non-violent trauma of climate induced migration will be very different, being that climate change is not as tangible or as easy to personify as a man in a plane dropping bombs overhead, just as is seen in a warzone there will be land that can no longer be walked upon, and communities lost in the chaos. The needs of those who are displaced by climate change in terms of education, medical care, accommodation, jobs, and access to resources will all present their own challenges according to age, gender, sexual orientation, ethnolinguistic backgrounds, etc. In the case of Indo-Fijians we must utilize the collectivism and strength that Indian diaspora groups possess, which has historically enabled them to survive and thrive outside of their traditional homeland. That same resilience will hopefully help to strengthen their resolve and improve their climate resilience methods and infrastructure which could slow migration levels and produce an inclusive society which values the presence and contribution of Indo-Fijians.

THE FUTURE

Secondary migration (Kim 2004) appears to be the most sensible option for many Indo-Fijians, particularly those who already have family ties beyond their borders or who have the financial means to do so. But what of those who do not fall into either bracket or would prefer to stay in a place they now call home? Newer and more inclusive methods of climate resilience will need to be implemented across Fiji to ensure that the lives and livelihoods of Indo-Fijians are protected. Strengthening community ties across ethnic lines through education, and recognition of their reliance on one another, will be of major significance if this becomes their chosen path. Indo-Fijians will need to be welcomed into the Vanua of Fiji's various communities in order to bypass the regulations that inhibit their access to natural resources and to allow for TEK (Traditional Ecological Knowledge) to be passed through the wider community. Failing to do so will result in the loss of a significant proportion of the Fijian population, which will also affect the government's financial ability to build climate-resilient infrastructure and to pay for damages to homes, business, and communal amenities. The World Bank estimates that the replacement of assets currently at risk in Fiji will cost $22,175 million dollars and annual average economic costs will reach 2.6% of GDP or $79.1 million dollars (Acting on Climate Change & Disaster Risk for the Pacific 2013). Both indigenous Fijians and Indo-Fijians will have to shoulder this tremendous financial burden. If they are unable to work collectively on this, Fiji will fail in the rebuilding of crucial infrastructure after extreme weather events, and with decreased numbers, those Indo-Fijians who decide against migration will have less say in which climate resilience projects receive investment and which communities will benefit from post and pre-disaster aid.

The Fijian government has set out key opportunities for mitigation across various sectors in their 2012 National Climate Change Policy. In farming, suggestions have been made for the "use of farming practices that maintain or increase forest cover (agroforestry), ensuring minimal soil tillage and soil cover to prevent the release of carbon in soil, reducing the use of fertilisers that can be converted and released as greenhouse gases, intensification of small-scale commercial and subsistence agricultural activities to optimise production which can minimise forest clearance and capturing methane gas from manure" (Republic of Fiji National Climate Change Policy 2012). Such suggestions rely on traditional knowledge for agroforestry and integrated farming methods, the applications of which will require the organization and unification of the Fijian populace. As change must occur across all sectors of society related to agriculture, they cannot afford to have indigenous farmers purchasing substandard fertilizers from Indo-Fijian businesses or indigenous landowners instructing Indo-Fijian tenant farmers to plant crops on farmland which is not or will not be productive in the years to come. The diversity of the Fijian population should be beneficial if one group can learn from another and if they work together towards using more sustainable farming methods, which will build up resilience for both groups whilst buying time for the international community to take significant steps to slow the rate of climate change.

Fishing also cropped up in the Fijian governments 2012 policies; however, they gave no examples of mitigation opportunities, having only recognized the impacts of climate change as well as a couple of properties which contribute to climate change resilience such as "mangrove areas and coral reefs and other coastal zones providing physical buffers to extreme weather events" as well as having "healthy reef ecosystems which are more resilient to the impacts of climate change, such as ocean acidification and increasing sea water temperature" (Republic of Fiji National Climate Change Policy 2012). Naturally fishermen/women across the Fijian islands will look to the government, elders, and scientists to provide methods of resilience. Yet, Indo-Fijians have been fishing Pacific waters now for generations and have built up an understanding of how climate change and pollution are affecting their livelihoods. If Indo-Fijians are included by being given equal fishing rights they will not only benefit financially but will have a say in how fishing is conducted whilst providing knowledge and experience which can be applied to climate mitigation systems. Fishing for Indo-Fijians can be compared to the difference between renting and buying a home. If you rent then you are less concerned by that scuff in the carpet, the mark on the wall, or the water damage in the bathroom as nothing belongs to you. However, if you buy a home then you are more careful and will fix any damages as quickly as possible. You will care more about the place in which you reside because you feel as though you are a part of it and that you will be directly affected if anything were to happen. This is not to say that Indo-Fijian fishermen/women are not concerned by the changes in the Pacific Ocean as this is clear from our earlier discussion. Having said that, the exclusion of numerous citizens who rely on fishing as a means of making a living will have detrimental impacts upon indigenous and Indo-Fijians alike and we must ask ourselves, at this current time, why Indo-Fijians would take the time and effort to contribute to a sector which could shut them out at any moment?

Indo-Fijian homes and businesses face somewhat different challenges to the fishing or agricultural sectors. Nevertheless, numerous ideas have been proposed to slow Fiji's contribution to climate change, e.g. "increased energy efficiency and use of renewable energy in residential, commercial and industrial sectors, reduction of household waste burning, promotion of household composting, including use of compost toilets, improvements to landfill management and increased recycling facilities and collections" (Republic of Fiji National Climate Change Policy 2012). As Indo-Fijians make up a large proportion of the urban population their participation in eco-friendly activities will be vital to the future carbon neutrality of Fiji. By helping to limit their contribution to climate change there may be less immediate examples of anthropogenic climate change and pollution caused by the residents of the area. Indo-Fijians can in a sense protect themselves by taking actions to change the way they live their lives. They will also need to look to their indigenous counterparts when extreme weather events occur which destroy businesses, as some traditional indigenous building practices provide more resilience to such events. By rebuilding and investing in traditional knowledge, Indo-Fijians and indigenous Fijians can work together, not only through the post disaster recovery period, but by building long-term connections with one another and relaying lived experiences, which reduce the ethnic tensions that lead to inequality and infighting.

Concluding Thoughts

To speculate upon the future outcomes of climate change may appear foolish, but if we allow ourselves to keep moving forward to the point where futile attempts at mitigation and adaptation leave minority ethnic groups vulnerable to political scapegoating and possible violence, we would be making a far more irresponsible choice and could be responsible for devastating future consequences. By recognizing how Indo-Fijian communities will cope with disasters and crises and by listening to affected people to discover what aid they want and need, we can start to lay plans for achieving either migration with dignity or moulding the social cohesion Fiji needs to survive the threats climate change poses.

If migration is the chosen technique for coping with climate change then it will be of great significance that we listen to advice from the OECD in ensuring that Indo-Fijian migrants who are displaced or choose to move are not considered as part of a homogeneous group of migrants (Ramos et al. 2018), with all the negative connotations that entails. We must not let the discourse of public arguments on migration, which often lack the credibility of using accurate statistics, hinder the upwards mobility of Indo-Fijian migrants in their newly chosen homes. Though Australia and New Zealand have developed pre-arrival assessments of foreign qualifications to select migrants (Ramos et al. 2018), which should make for an easier transition for educated Indo-Fijians when choosing to migrate. Stability still cannot be ensured for all, particularly for those who will be displaced by climate change as their movements may be slow or sudden depending on the circumstances, thereby affecting education, job prospects, not to mention healthcare, gender inequalities and a loss of culture and language. We must also consider those Indo-Fijians who do not possess the relevant qualifications or language skills which are necessary for starting a new life in a foreign country. These factors may hinder their prospects for attaining citizenship or a visa and leave them in a difficult form of legal purgatory which could lead to other more significant issues down the line. For both groups the implementation of integration policies will be key to promoting growth and social cohesion (Ramos et al. 2018). There will need to be a concerted effort made on both sides to ensure all migrants are given the opportunity to contribute their skills and potential to their new nations.

For those Indo-Fijians who choose to remain, the policies put in place by the Fijian government to combat the impacts of climate change will need to recognize the non-homogeneous group of people that make up the Fijian nation. This means planning for the movement of Indo-Fijian communities who live on a shoreline at risk of inundation, enforcing the law to ensure equality and to stop any discrimination towards Indo-Fijians, in particular for those who require insurance against climate induced disasters for their businesses and livelihoods and most importantly finding new methods to integrate both communities and heal historical divides. It is everyone's responsibility in Fiji to change their mind-sets, otherwise they won't endure the inevitable disasters to come. However, Fiji can only do so much to limit its impacts on the natural world. The governments and residents of the global north and developing countries which are contributing the most to climate change need to start making changes today. This includes reforming our views on migrants and educating ourselves on the differentiated consequences of climate change faced by minority groups across the Pacific Islands and beyond. International law and policy should reflect this dynamic and have frameworks available which are flexible enough to provide effective relief to all threatened groups who find themselves in a system of discrimination or inequality. Industries, businesses, governments, and private citizens cannot continue to fill their back pockets whilst ignoring the most dangerous threat to humanity we have or will ever see. Around 200 million people, which is the equivalent of the entire Nigerian population (the seventh most populated nation on earth) (Total Population by Country 2021 2021) could be displaced by climate change by 2050 (Brown 2008) and this is a prediction we should all be worried about, whilst working to reverse or at least slow it down. Dependent upon where we live we may not be directly affected today, tomorrow, or even in five years' time but there will come a day when we all feel the force of climate change and by then it will be too late. We need to "Wake Up!" (Porter 2018).

REFERENCES

Acting on Climate Change & Disaster Risk for the Pacific [ebook]. 2013. The World Bank, pp. 6, 7. Available at: https://www.worldbank.org. Acting on Climate Change & Disaster Risk for the Pacific - World Bank Group.

Baird, R., 2008. The Impact of Climate Change on Minorities and Indigenous Peoples [ebook]. London: Minority Rights Group International. Available at: https://minorityrights.org. The Impact of Climate Change on Minorities and Indigenous Peoples.

Balachander, V., 2012. Sunday ET: Pickle Making, an Age-Old Ritual Still Prevalent in Indian Households [online]. *The Economic Times*. Available at: https://m.economictimes.com/sunday-et-pickle-making-an-age-old-ritual-still-prevalent-in-indian-households/articleshow/15547764.cms.

Barnett, J., 2005. Titanic States? Impacts and Responses to Climate Change in the Pacific Islands. *Journal of International Affairs*, 59(1), 203–219. Available at: http://search.proquest.com/docview/220715483/.

Batman, G., 2015. Is Edward Said's Orientalism Relevant for the Pacific? [online]. Geopolitical Grapplings. Available at: https://nicholasanda.wordpress.com/2015/08/13/is-edward-saids-orientalism-relevant-for-the-pacific/.

BBC. 2020. Climate Change: Rising Sea Levels in Fiji Create 'Ghost Towns' [video]. Available at: https://www.bbc.co.uk/news/av/world-asia-54138677.

Brown, O., 2008. Migration and Climate Change [ebook]. Geneva, Switzerland: IOM International Organisation for Migration. Available at: https://www.ipcc.ch. Web results Migration and Climate Change – IPCC.

Bryant-Tokalau, J., 2018. *Indigenous Pacific Approaches to Climate Change*. Dunedin, New Zealand: Palgrave.

Campbell, John R., 2014. Climate-Change Migration in the Pacific. *The Contemporary Pacific*, 26(1), 1–28. http://www.jstor.org/stable/23725565.

CCTV. 2013. Fiji Sugar Farmers Struggle [video]. Available at: https://youtu.be/iRTqENi_EpI.

Cox, M., Arnold, G., and Tomás, S. V., 2010. A Review of Design Principles for Community-Based Natural Resource Management. *Ecology and Society*, 15(4).

D'Souza, E., 2001. Indian Indentured Labour in Fiji [ebook]. Indian History Congress. Available at: https://www.jstor.org/stable/44144422.

Daye, Russell. 2009. Poverty, Race Relations, and the Practices of International Business: A Study of Fiji. *Journal of Business Ethics*, 89, Springer, 115–127. http://www.jstor.org/stable/27749762.

Devadason, C., Jackson, L., and Cole, J., 2019. Pacific Island Countries: An Early Warning of Climate Change Impacts [ebook]. Oxford: The Rockerfeller Foundation. Available at: https://www.planetaryhealth.ox.ac.uk. Pacific Island Countries: An Early Warning of Climate Change Impacts.

Effects of Climate Change on Human Mobility in the Pacific and Possible Impact on Canada [ebook]. 2016. Canberra: International Organisation for Migration, pp. 17–18. Available at: https://publications.iom.int. IOM Publications - International Organization for Migration (Effects of Climate Change on Human Mobility in the Pacific and Possible Impact on Canada).

Everyculture.com. 2004. Indo-Fijians - Introduction, Location, Language, Folklore, Religion, Major Holidays, Rites of Passage [online]. Available at: https://www.everyculture.com/wc/Costa-Rica-to-Georgia/Indo-Fijians.html.

Firth, S. (Ed.), 2006. *Globalization and Governance in the Pacific Islands: State, Society and Governance in Melanesia*. ANU Press. Accessed June 2, 2020, from www.jstor.org/stable/j.ctt2jbj6w.

FX, 2018, Porter. B, Pose. https://youtu.be/cE3fijgMzos.

Hollup, O., 2000. Structural and Sociocultural Constraints for User-Group Participation in Fisheries Management in Mauritius. *Marine Policy*, 24(5), 407–421.

https://fijisun.com.fj/2008/06/21/the-life-of-a-fisherman-a-socioeconomic-survey/.

Keener, V., Marra, J. J., Finucane, M. L., Smith, M. H., and Spooner, D. (eds.). 2013. Climate Change and Pacific Islands: Indicators and Impacts: Report for the 2012 Pacific Islands Regional Climate Assessment. Washington: Island Press. Available from: ProQuest Ebook Central [2 June 2020].

Kim, Kyung-hak. 2004. Twice Migrant Indo-Fijian Community in Sydney with Particular Reference to Socio-Religious Organizations. *Indian Anthropologist*, 34(2), 1–27. http://www.jstor.org/stable/41919963.

Kumar, Mukesh. 2012. A Quest for Identity/Equality: Indians in Fiji, 1879–1970. *Proceedings of the Indian History Congress*, 73, 1053–1064. www.jstor.org/stable/44156305.

Kumar, Nirmalya, E. M. Steencamp, Jan-Benedict. 2013. *Harvard Business Review*. Diaspora Marketing [online]. Available at: https://hbr.org/2013/10/diaspora-marketing.

Kurer, Oskar. 2001. Land and Politics in Fiji: Of Failed Land Reforms and Coups. *The Journal of Pacific History*, 36(3), 299–315. http://www.jstor.org/stable/25169559.

Lal, Brij V. 2012. *Chalo Jahaji: On a Journey Through Indenture in Fiji*. ANU Press http://www.jstor.org/stable/j.ctt24h3ss.

Lal, Brij V., and Pretes, Michael (eds.). 2008. *Coup: Reflections on the Political Crisis in Fiji*. ANU Press. http://www.jstor.org/stable/j.ctt24h2nd.

Leonard, P., 2019. Land and Belonging in An Indo-Fijian Rural Settlement [ebook]. Manchester: The University of Manchester. Available at: https://www.research.manchester.ac.uk. Land and Belonging in an Indo-Fijian Rural Settlement - Research Explorer.

Manghubai, S., Nand, Y., Reddy, C., and Jagadish, A., 2021. Politics of vulnerability: Impacts of COVID-19 and Cyclone Harold on Indo-Fijians Engaged in Small-Scale Fisheries [ebook]. Sciencedirect. Available at: https://www.sciencedirect.com/science/article/pii/S1462901121000745

Minority Rights Group. 2017. Indo-Fijians - Minority Rights Group [online]. Available at: https://minorityrights.org/minorities/indo-fijians/.

Morrison, R. J., et al. 2013. Anthropogenic Biogeochemical Impacts on Coral Reefs in the Pacific Islands-An Overview. *Deep-Sea Research Part II: Topical Studies in Oceanography*, 96(C), 5–12. https://doi.org/10.1016/j.dsr2.2013.02.014.

Narang, S., 2018. After Devastating Cyclone, Fiji Farmers Plant for a Changed Climate [online]. Npr.org. Available at: https://www.npr.org/sections/thesalt/2018/01/09/573521139/after-devastating-cyclone-fiji-farmers-plant-for-a-changed-climate.

Neef, A., 2019. Climate Change Adaptation In Post- Disaster Recovery - Policy Brief 7 Is Planned Relocation a Viable Solution to Climate Change Adaptation Policy in Fiji? [ebook] University of Auckland, p. 3. Available at: http://www.climatechangeadaptation. In Post-Disaster Recovery - Policy Brief 7 Is

Planned Relocation a Viable Solution to Climate Change Adaptation Policy in Fiji?

Oliver-Smith, Anthony. 1996. "Anthropological Research on Hazards and Disasters." *Annual Review of Anthropology*, 25, 303–328, http://www.jstor.org/stable/2155829.

OECD Education and Skills Today. Available at: https://oecdedutoday.com/why-we-should-dispel-the-myth-of-migrants-as-a-homogeneous-group/.

Ramos, G., Schucknecht, L., Scheicher, A., and Scarpetta, S., 2018. Why We Should Dispel the Myth of Migrants as a Homogeneous Group. OECD Education and Skills Today [online].

Reddy, C., 2018. Hidden Figures: The Role of Indo-Fijian Women in Coastal Fisheries [ebook]. Available at: https://www.spc.int. WIFPDF The Role of Indo-Fijian Women in Coastal Fisheries - The Pacific Community.

Reddy, C., 2019. Indo-Fijian Fishing Communities: Relationships with Taukei in Coastal Fisheries [ebook]. Victoria University of Wellington. Available at: https://researcharchive.vuw.ac.nz/xmlui/bitstream/handle/10063/8878/thesis_access.pdf?sequence=1. Indo-Fijian Fishing Communities: Relationships with Taukei in Coastal Fisheries - VUW Research Archive.

Reddy, Mahendra. 2007. Small Business in Small Economies: Constraints and Opportunities for Growth. *Social and Economic Studies*, 56(1/2), 304–321. http://www.jstor.org/stable/27866505.

Refugees, U., 2017. Refworld | World Directory of Minorities and Indigenous Peoples - Fiji Islands [online]. Refworld. Available at: https://www.refworld.org/docid/4954ce3e53.html.

Republic of Fiji National Climate Change Policy [ebook]. 2012. Suva, Fiji: Secretariat of the Pacific Community, pp. 9, 10, 11, 12, 13. Available at: https://www.sprep.org. Republic of Fiji – National Climate Change Policy – SPREP.

Sherif, S., 2013. Negotiating Postcolonial Spaces: A Study of Indo–Sri Lankan Fishing Disputes. *International Studies*, 50(1–2), 145–164.

Singh, J., 2008. The Life of a Fisherman: A Socio-Economic Survey. Retrieved from.

Sleet, P., 2019. Fiji: Poor Nutrition and Agricultural Decline Has Caused Food Security Slump - Future Directions International [online]. Future Directions International. Available at: https://www.futuredirections.org.au/publication/fiji-poor-nutrition-and-agricultural-decline-has-caused-food-security-slump/.

Sundaresan, T., 1985. Food Plants in Fiji and Their Utilization [ebook]. Kagoshima University, pp. 146–159. Available at: 2016. Effects of Climate Change on Human Mobility in the Pacific and Possible Impact on Canada [ebook]. Canberra: International Organisation for Migration, pp. 17–18. Available at: https://publications.iom.int. IOM Publications - International Organization for Migration.

Tegunimataka, A., and Palacio, A., 2021. Oceans Apart - Internal Migration in a Small Island Developing State: The Case of Fiji [ebook]. Lund, Sweden: Lund University. Available at: https://www.ed.lu.se/papersPDF Oceans Apart - Internal Migration in a Small Island Developing State: the case of Fiji.

The Pacific Way Story - Struggling for a Better Life, Squatters in Fiji. 2014. Fiji: Pacific Community. https://youtu.be/nkTJ220Bb-Q.

Trnka, Susanna. 2005. Land, Life and Labour: Indo-Fijian Claims to Citizenship in a Changing Fiji. *Oceania*, 75(4), 354–367. http://www.jstor.org/stable/40331994.

Trnka, Susanna. 2008. *State of Suffering: Political Violence and Community Survival in Fiji*, 1st ed. Cornell University Press. http://www.jstor.org/stable/10.7591/j.ctt7zhr1.

Worldpopulationreview.com. 2021. Total Population by Country 2021 [online]. Available at: https://worldpopulationreview.com/countries.

Exploring Australia and New Zealand's Climate Policies: Similarities and Differences

Aarushi and Pavan Kumar

Australia and New Zealand are neighbours in the Pacific region. The activities and policies of each have an impact on each other as well as on other countries in the region. Australia for long has enjoyed its economic prosperity through its reserves of coal and other fossil fuels. It is one of the largest exporters of coal (IEA 2018). Thus, it generates enough carbon emissions that pollute the environment. It was ranked last on the Climate Change Performance Index out of 57 countries, which are responsible for more than 90% of the emission of greenhouse gases (CCPI 2020). Australia has set up a target of 26–28% reduction in its carbon emissions by 2030, under the Paris Agreement (Goodman 2020). However, many countries are not happy with this target and claim that it is unambitious. The New Zealand government has passed the Zero Carbon Act 2019 after signing the Paris Agreement. The Climate Change

Aarushi
University of Delhi, Delhi, India

P. Kumar (✉)
Janki Devi Memorial College, University of Delhi, Delhi, India
e-mail: pavan.sis.jnu@gmail.com

© The Author(s), under exclusive license to Springer Nature Switzerland AG 2022
N. J. P. Alsford (ed.), *Pacific Voices and Climate Change*,
https://doi.org/10.1007/978-3-030-98460-1_10

Commission of the country is responsible for meeting the target of net-zero emissions by 2050. The government of the country further aspires to become carbon neutral by 2025 (Ministry of Environment 2021b). The country also declared a climate change emergency last year in November. However, many voices from the opposition party said that this is just a "virtue signalling" and no concrete step will be taken further. Moreover, the targets set by the government are far below what the country should do (Taylor 2020).

Both the countries' domestic activities have an impact on other countries in their region and so do their policies related to climate. Both the countries boast of having climate policies that will contribute significantly to the environment. However, both these countries have been targeted for insufficient policy adoption and improper execution. Some stakeholders affect the decision-making and policies of these states and have been continuing to do so even after decades of debates on climate change and its impact. The international pressure on the countries also affect the development of policies in the two countries. The pressure is both in economic and political terms so as to make the countries responsible for their actions towards the environment.

Thus, to study the climate change responses of the two countries, this paper has used the reports on the climate change actions taken by both the government of Australia and New Zealand. Apart from the primary reports, it has also used secondary literature on the issues to interpret and analyse the policies of both the states. This piece of literature was extensively used to interpret and analyse the various kinds of pressure that these countries face during their policy formation on climate change. Thus, using interpretative approach as a method.

To keep the issues in order, this paper is divided into four parts. The first two parts briefly talk about the climate policies adopted by Australia and New Zealand, respectively, from the 1980s onwards. Thereby giving a summary of the policies adopted till now and the role of different organizations in influencing the climate policy formulation in the two countries. The third section talks about the reasons and the people who affect the policies, their introduction, execution, and denial. It talks about the role of both domestic and international groups that affect the policy formulation in any country. The fourth part traces the similarities and differences in the climate policies of Australia and New Zealand thus providing a brief overview of the situations in the two countries.

POLICIES ADOPTED BY AUSTRALIA

Australia's engagement with climate change starts in the 1980s. A community of scientists and researchers were studying the impacts of global climate change on Australia and the way the country is account-able for the same (Bulkeley 2000). These analyses were happening each at the domestic level and the international level. The Villach Conference of 1985 delineates an effort to review this research under the auspices of the World Meteorological Organization, the International Council of Scien-tific Unions, and the United Nations Environment Programme (Taplin 1996). This conference brought the problem of climate change into the political landscape of the country. It asked the stakeholders to use policy measures to address the issue.

The government research organization, CSIRO (Commonwealth Scientific and Industrial Research Organization), took interest in conducting studies on carbon dioxide concentration and the effect of greenhouse gases on the environment. It played a crucial role in imple-menting the outcomes of the Villach conference into political develop-ments. CSIRO along with the Commission for the Future (CFF), an organization established by the Federal government to boost the extent of debate over key scientific and technical developments, CSIRO orga-nized two Greenhouse "information" events in the late 1980s. The first, Greenhouse'87, was an effort to interact with the interest of the scientific community, assess what the possible impacts of climate change could be, and draw in political and public attention. The second event, Greenhouse'88, created a public forum for the discussion of Greenhouse problems. These events won CSIRO and CFF a 'Global 500' award from the United Nations and created an enduring impression on both policymakers and the public (Bulkeley 2000: 37).

While this was happening at the domestic level in Australia, the globe was heading towards a significant conference at Toronto in the year 1988, on the subject "The Changing Atmosphere". The freshly and extensively debated topic of global warming was placed at the centre of the table here. This conference urged the government to create an international advisory body through which it will address the difficulty of climate change and thereby the policy responses relating to the same. The Intergovernmental Panel on Climate Change (IPCC) was established in 1988 to provide policymakers with regular scientific assessments on the present state of information regarding climate change (IPCC 2021). Further, "Toronto Target" was conjointly adopted.

These international pressures and also the domestic pressures from Australia's scientific community led the Federal Government to act on the difficulty of greenhouse emissions. Thus, in April 1989 the Federal Government set up a national climate change program, which included the establishment of the National Greenhouse Advisory Council (NGAC). Throughout 1990, the Australia and New Zealand environment Council emphasized the necessity for a National Greenhouse strategy and the Federal Government took the primary steps towards that end (Bulkeley 2000: 37). The state governments conjointly adopted numerous versions of the Toronto Target.

After the release of the first IPCC report, the international pressure on the Federal government grew considerably. The pressure mounted as the world was heading towards the Second World Climate Conference. On 11 October 1990, just before the international conference, the Federal Government adopted the "Interim Planning Target" (IPT) (Bulkeley 2000: 38). This target was in the lines of the Toronto conference. It committed the government to reduce emissions of greenhouse gases to 1988 levels by 1990 and cut emissions by 20% by 2005. This target made sure that in doing so the Australian economy would not be affected and there would be no impacts on its trade competitiveness.

At the national level, the federal government started working on a national environmental strategy which was guided by the principle of Ecologically Sustainable Development (ESD). The ESD method described Australia's most cooperative plan to type a broad environmental strategy and to flee this endemic environment-economy, intergovernmental conflicts (Kinrade 1995; Dowens 1996). The Federal government played a key role in selecting the participants to ESD Working Group, selecting them from what it deemed "key interest organizations", and thereby encouraging the "monopolistic representation" of interests (Bulkeley 2000: 40). The ESD working group was asked to prepare a report for cost-effectively meeting the IPT. It found several cost-efficient actions to reducing greenhouse emissions.

The National Strategy for Ecologically Sustainable Development (NSESD) and a National Greenhouse Response Strategy (NGRS) were two reports drafted from the reports of the working groups. These documents excluded antecedently written negotiations in their attempt to seek out policy choices and bore few similarities to the first document. Meanwhile, at the domestic level, the Industry Commission was asked to report on the cost and benefits of stabilizing greenhouse gas emissions for the Australian industry. It reported that the national output would cut back by 1.5% if IPT is followed. Further, though most sectors of the economy would be affected the cost would be borne more by certain sectors and regions. The report also stated that the cost of inaction calculation is difficult as there is uncertainty surrounding the impact of greenhouse gas. Further, the role of the carbon tax was also introduced (Industry Commission Annual Report 1990).

In 1992, Australia joined other nation-states in Rio to sign the Framework Convention on Climate Change (FCCC). However, the federal government displayed an unwillingness to accept the legally binding targets and timetables due to the pressure from the industrial sector. It, therefore, advocated the differentiated approach (Bulkeley 2000: 33). From this time onwards, Australia's stance on climate change subsequently became questionable. The country adopted the "no regrets" which refers to the adoption of such measures that have low social and economic costs and meet the requirement without hampering the economy and the output. This stand took a deviation from the ESD Working Group report.

Further, the pursuits of the IPT and NGRS were given up by not committing towards them. The government did not have any defined targets and deadlines towards its goals and did not commit itself to any new goals. The government responses were ad hoc and left to the mercy of commercial decisions. The representatives of the large and well-organized industries attracted the support of the government. The finance and development departments of the state sympathized with the "economic rationalism" approaches of these interest groups (Bulkeley 2000: 43). The government also cut spending towards many of its programs.

With renewed international developments surrounding the design of a climate "protocol" and the impending release of the IPCC's second assessment report, placed Greenhouse concerns back on the domestic agenda during 1994. The possibility of implementing a carbon tax was considered but was met with concerted industry opposition, and instead, they proposed the "no regret" measures. The "Greenhouse Challenge" Programme (GCP) was introduced by the federal government. It encouraged companies and industry associations to sign up for voluntary reductions in emissions (International Energy Agency 2017). The GCP was complex and its objectives were unclear. In early 1995, a supplement was formulated to the 1992 NGRS, Greenhouse 21C: A Plan of Action for a Sustainable Future (Department of Environment, Sport and Territories 1995). This was Australia's domestic climate policy response for the Berlin Conference of the Parties.

Australia's position in the lead-up to Kyoto was to favour differential targets and to oppose any uniform reductions in emissions. It was argued that since Australia is heavily dependent on fossil fuels for domestic energy and export revenue, thus it would be costly for the country if uniform targets would be imposed. Thus, Australia advocated differentiation that is the allotment of different targets for different countries based on the economic costs due to reductions of each country. This approach of the government was problematic and many Australians—including scientists, economists, energy experts, and ordinary citizens—protested at what they saw as its short-sighted and self-defeating stance. In June 1997, a statement signed by 131 professional economists, together with sixteen full professors of economics, called on the Government to reverse its position (Hamilton 2000: 52). Moreover, the Australian clause of land clearing in the Kyoto protocol was a loophole and gave an unfair advantage to the country. Thus, Australia took advantage of the more responsible approach adopted by other countries and exploited the fact that agreement on mandatory targets by all Annex I countries were essential to obtaining a protocol.

Post Kyoto Australia had some major initiatives in the direction of climate change. Post-2000 Australia has shown more consistency on domestic climate policy than ever before but it was less consistent on the climate diplomacy front (Beeson and Mc Donald 2013: 335). It focused more on the pricing and trading of carbon emissions. Australia's national elections in 2007 saw a heated debate and discussion around Australia's climate policy. While Howard was opposed to ratifying the Kyoto protocol, but Rudd was keener on agreeing to some kind of carbon emission policy. Labour Party leader Kevin Rudd in the course of the campaign declared that he would sign the Kyoto pact if voted in power. He said, "There is no better way to reinforce that than prime ministerial attendance. It would be a way of indicating ... that we intend to be globally, diplomatically active" (Taylor 2007). His contender Howard said, "The world is not coming to an end tomorrow. Like all of these things we have to get a common sense, balanced approach. You need a new international agreement that includes countries like China" (Taylor

2007). Thus, post winning the elections Kevin Rudd signed the Kyoto pact in 2007 (NBC 2007). But as Tim Johnston argues in 2007, for Rudd, the hard part came after signing the pact (Johnston 2007). From 2000 to 2005, Australia was interested only in climate science, voluntarily advocating energy efficiency and investing in non-renewable energy resources. But they did not focus on reducing the carbon emission at home, that would impact the industry as well as export. But after the new government in 2007, Australia's domestic climate policies targeted carbon pricing and stressed the adaptation approach and a holistic and integrated policy. And this led to the Clean Energy Package of 2011. Kate Crowley believes this to be the most important step from the government in Australia's climate policy history (Beeson and Mc Donald 2013: 336). Australis's Department of Climate Change and Energy Efficiency had four goals of its climate policy in 2012. And these were reducing greenhouse gas emissions; promoting energy efficiency; adapting to climate change impacts; and helping to shape a global solution (DCCEF 2012). If the first two aimed at mitigation, the third focused on adaptation and the fourth one was to emphasize climate diplomacy (Beeson and Mc Donald 2013: 336). Australian government further targeted a 5% reduction of greenhouse gases by the year 2020. With these, it is also expected to meet its 20% energy requirement by renewable energy sources. This was Australia's commitment in 2012 (National Communication on Climate Change, NCCC 2013). Minister of Climate Change, Mark Butler writes in the forward, "The Australian Government has committed to responsible targets to reduce carbon pollution and to play our part in the global effort to avoid dangerous climate change. The Government has committed to reducing carbon pollution by 5% from 2000 levels by 2020 irrespective of what other countries do, and by up to 15 or 25% depending on the scale of global action. Looking out to 2050, the Government has committed to cut carbon pollution by 80%below 2000 levels" (NCCC 2013: 1). Thus, it was a big event in Australia's domestic politics. It aimed at portraying Australia as a responsible actor in international politics. And it designed some of its policies to do the same. To highlight its commitment, it advertised its progress between the 5th and 6th communication on climate change. The 6th communication stated,

Key Achievements Since Australia's Fifth National Communication On Climate Change:

- Commitment to ambitious emissions reduction targets, including the adoption of an 80% reduction in emissions on 2000 levels by 2050;
- Development of a comprehensive Clean Energy Future Plan to address the challenge of climate change;
- Introduction of an economy-wide carbon price in 2012;
- Passage of legislation to establish the Clean Energy Regulator, Climate Change Authority and Clean Energy Finance Corporation;
- Creation of an independent Climate Commission staffed by scientists, Economists, and public policy specialists to provide accessible information on climate change to the general public;
- Development of a Plan for Implementing Climate Change Science and a National Climate Projections Program to further Australia's climate change science research and observation efforts;
- Increased understanding of the impacts of climate change on Australian ecosystems and our coastline; and
- Continued and enhanced investment in adaptation, mitigation, capacity building and technology cooperation throughout the Asia-Pacific region. (NCCC 2013: 30).

The government founded a $126 million Climate Change Adaptation Program. This was aimed at its adaptation strategies. National Climate Resilience and Adaptation Strategy were formulated in 2015 where these details were made public. This was a step ahead of the National Climate Change Adaptation framework 2007. The strategy document claimed that Australia is interested in making a set of principles that would guide effective adaptation practices and resilience building (NCRAS 2015). It further connected its climate policy with its $13 billion Secure Water Supply program. It was an effort to show the commitment to help the Australian farmers who were affected negatively by changing climate patterns such as desertification, rainfall, floods, etc. (Beesan and Mc Donald 2013: 336).

In 2015, Liberal Government announced Australia's intention to ratify the second commitment period of the Kyoto Protocol. Australia joined the Paris Agreement which aimed at keeping the global warming "well below" 2 °C. But at the same time, Australia's emissions rose in 2015 (Talberg et al. 2016).

Currently, Australia's climate change policies are under the Department of Industry, Sciences, Energy, and Resources. On its website, it highlights its climate change strategies. It talks about domestic climate policy and changes it has done to deal with climate change. It says that the department aims to chart policies that are aimed at meeting its obligation under the Paris Agreement and to do that it has to reduce its greenhouse gas emission (DISER 2021). It further talks about Australia's domestic policy on climate change. Domestic policies focus on climate change programs that help reduce emissions, develop and coordinate renewable energy policy, interact and interact with different stakeholders, help and promote businesses and industries to have smart technology and practice, and finally to help the people in the agriculture sector to reduce the carbon emission. And all of these responsibilities lie with the Department of Agriculture, Water and the Environment (DISER 2021). It reiterates its international climate change commitments. It says that Australia is the party to the Paris agreement which came into force in 2016 and it also respects UNFCC and Kyoto Protocol. It states that Australia, under the Paris Agreement, submits emission reduction commitments. These commitments are known as Nationally Determined Contributions (NDC). And its first NDC in 2015 had a target of reducing greenhouse emissions by 26 to 28% below 2005 levels by 2030. In 2020, it further recommunicated its 2030 targets. The next NDC is due in 2025. And to achieve these targets, the concerned department is leading the negotiations on land sectors., reporting emissions, supporting the developing countries by adding in projects, to look after and approve the participation in Clean Development Mechanism and Joint Implementations project. It has also signed an MoU with Singapore to advance cooperation in technologies that help in low carbon emission (ICCC 2021).

But international observers and agencies do not appreciate Australia's commitments and consider them only as a face-saver. They argue that Australia is not doing much on climate change issues. In 2019, United Nations criticized Australia for having too low goals and it said that Australia is not on track. It said, "There has been no improvement in Australia's climate policy since 2017 and emission levels for 2030 are projected to be well above the target" (Goodman 2020). The Climate Change Performance Index ranked 54 out of 61 countries. This was ranked as an overall very low performance (CCPI 2021. The report further says, "The country receives very low ratings in three of the four CCPI categories: GHG Emissions, Energy Use, and Climate Policy, and a low rating in Renewable Energy. Despite positive tendencies in the trend indicators, with a growing share of Renewable Energy in Energy Use and an overall decrease in the per capita Energy Use, as well as in the per capita GHG emissions indicator, current levels and future targets across all categories are not on track with a well-below -2 °C pathway. Together with the United States, Australia holds the last place for its climate policy evaluation" (CCPI 2021). In 2020, Australia was excluded from Global Climate Talks as it lags in its policies and talks on climate change (Scott and Jess 2020). Scientists have also criticized Australia's Aus$3.5 billion plan of climate change as this would not encourage the industries to reduce the emission (Nogarady 2019). According to a projection, Australia's emissions would be only 16% lower than 2005 levels in 2030. Australia's coal industry is also allegedly contributing to the carbon emission, thus Australia is indirectly responsible for the pollution. Australia was the fourth-largest producer in the coal industry in the year 2017, and it is crucial that they cut the coal industry (Goodman 2020).

POLICIES ADOPTED BY NEW ZEALAND

New Zealand was the home of the world's first national "green" party—the Values Party, in 1972 (Kelly 2010). It is a small country in the Pacific. It lies between the regionally dominant Australia and the small pacific island countries. It has many inhabitants of small island countries and therefore it is a supporter of the involvement of pacific island in the IPCC and FCCC processes.

The process of taking action towards climate change was started in 1988 with the establishment of the New Zealand Climate Change Programme (NZCCP) (Ministry for the Environment 2006). It was coordinated by the Ministry for the Environment. The country has a small scientific and research community and relationships that it has with small pacific island countries are the main reason behind this country's effort towards policy development and response to climate change.

In 1992, while the world was gearing up for the Framework Convention on Climate Change, the domestic voices for climate changes were disappearing slowly in New Zealand. By May of 1992, there were fears that the National government was about to abandon its 20% target in the lead-up to the Rio Earth Summit in June as economic impacts of policies were the foremost consideration. At this time, a 20% increase in emissions between 1990 and 2000 was reported. Later, when the government announced its Reduction Action Plan in July 1992, it was confirmed that there was no intention of adhering to its policy to "aim for" a 20% cut in emissions by the year 2000. No concrete actions were taken and the funds were also blocked. The climate actions were considered as a threat that may impact the government's policy of foreign direct investment (Hamilton 2000).

Thus, during this time a kind of apprehension towards domestic or national climate action was seen as a threat to the economy. It was felt that it would cost New Zealand economically and socially if other big countries do not take similar action towards climate change and emission reductions. Basher in the paper "The Impacts of Climate Change on New Zealand" cites three shortcomings that have led to the suggested loss. The first is the failure to develop and support an ongoing well-integrated climate change research program that follows up on the earlier impacts assessment work and provides systematic input to government policy-making and IPCC assessments. He says that the problem lies not with the structural arrangements of the existing science policy, purchaser, and provider organizations, but with the unduly ideological market modelled rules governing their relationships and in particular the reduced influence of policy departments over science resources. The second shortcoming, that he notes, is the failure to involve industry and business in science impacts research and assessment in a meaningful and constructive way, and the consequent ambivalence (and sometimes hostility) of these sectors towards climate change science and responses. The third shortcoming, failure to develop a national political consensus and will to undertake significant national no-regrets responses (both mitigation and adaptation) based on the science available (Basher 2000: 140).

Further, Kelly argues in his paper that three related factors—two political and one economic—have influenced climate policy in New Zealand to date. The relative strength of interest groups promoting and opposing meaningful action has been important. In terms of the economy, New Zealand has fallen behind its former peers in terms of national income. This created an urge to improve income, implying dependence on industries that have a significant impact on climate. Thus, stating the resemblance of New Zealand's position with other developing countries, facing the same conflict between economic development and sustainability (Kelly 2010).

Therefore, New Zealand's position on climate change policy till the mid-2000s reflects some missed opportunities whereby the policies could not mobilize the public towards environmental concerns. The government, further, did not dare to take the country towards the path of climate action as it constrained itself with the market approach. The country also failed in displaying moral leadership towards its fellow Pacific islanders by not introducing a new energy economy.

However, as the problem of climate change is increasing over time the country needs to react and take action accordingly in the present times. New Zealand is working towards developing its climate change policy to move towards living in a climate-resilient environment in the future. Its Climate Change Programme framework focuses on leadership at home and an international level. Thus, making it one of the few countries that have a net-zero emission target by 2050 written in its law. The framework also outlines a productive, sustainable, and climate-resilient economy along with a just and equitable society. This framework is in alignment with the country's commitment under the Paris Agreement and its target of reducing emissions by 30% below 2005 gross emissions for the period 2021–2030 (Ministry for the Environment 2021a).

The country also boasts of being on the path to low emissions and becoming a world leader in actions taken up to mitigate climate change. In this regard, New Zealand has been working towards climate change on two folds. One is its domestic level setting of targets and the other comprises of formulating policies to meet its international targets. This development is quite different from what we noticed in the early periods of policy formation. Wherein New Zealand was quite reluctant in formulating policies at the two levels.

The country introduced a new domestic emissions reduction target which it aspired to achieve by 2050 and it was set into law with the Zero Carbon Act in November 2019 (International Energy Agency 2021). It was an amendment to the Climate Change Response Act. This act was passed with broad party consensus. Thus, pointing to the fact that they need to take action on climate change has become increasingly important and urgent now. This broad party consensus also reflects the long-term nature of this act and the work towards the same.

Following this, a Climate Change Commission was also established in December 2019 (Climate Change Commission 2020). This commission is responsible for advising the government on climate change mitigation and adaptation. It is also responsible for helping the country progress towards making its new 2050 target emissions budget and the implementation of a National Adaptation Plan (Ministry for the Environment 2021a, b). The emission budget seeks to provide the total quantity of emissions that can be allowed to emit during a budget period. These emissions are calculated by looking at the sum of emissions in all the years across the target period rather than looking at each year individually. This emission measurement approach was mentioned under the Kyoto Protocol for the countries that agreed to meet the greenhouse gases reduction targets. Each of these budgets expands over 5 years. The first emission budget of New Zealand will be released in 2022 (Ministry for the Environment 2021a). These emission budgets thus act as "stepping stones" and will help New Zealand in achieving its long-term goals by specifying the targets each year. However, there are apprehensions about the Climate Change Commissions emission budgets as many believe that New Zealand will be making very less reduction in its emissions in the next ten years than the target of limiting warming to 1.5 degrees taken up at the Paris Agreement (Daalder 2021).

The government is also working towards an Emissions Reduction Plan (ERP) which will lay the path on which the country must move forward to meeting its 2050 target along with the emissions budget. New Zealand also runs an Emissions Trading Scheme (NZ ETS) to reduce emissions and meet its future reduction targets. It follows a market-based approach and charges certain sectors for their emissions. Some reforms were introduced in this scheme in the mid-2020 and the current scheme focuses on three areas particularly, market governance, industrial allocation reform, and technical regulation updates (Leining and Bruce-Brand 2020). It is one of the main schemes of the government to reduce emissions. However, since 2015 this scheme has turned to be a domestic scheme and the government is reluctant to use this for the international unit. The government also launched the Carbon Neutral Government Programme in December 2020, to make the public sector units carbon neutral starting from 2025 (Ardern et al. 2020).

On the international front, the country has adopted two broad targets over the period 2013–2020 and 2021–2030. The 2020 target was adopted under the UNFCCC and it seeks to reduce emissions by 5% below 1990 gross greenhouse gas levels for the period January 2013 to December 2020. The country says that it is about to meet this target. The 2030 target on the other hand seeks to reduce emissions 30% below 2005 gross emission for starting from January 2021 to December 2030. It is the country's first Nationally Determined Contribution (NDC) under the Paris Agreement (Ministry for the Environment 2021a). The updated NDC submitted in April 2020, many believe did not strengthen the country's 2030 target (Climate Action Tracker 2021b). The system and target set for the same are deeply flawed as the various positions are taken allow net emissions to increase over the period.

The country has adopted the net-zero emissions of all greenhouse gases other than biogenic methane by 2050 as a part of its domestic targets under the Climate Change Response (Zero Carbon) Amendment Act (CCRA) (Antonich 2020). Biogenic methane is the methane released from agriculture and waste. The emissions from the agricultural sector make up half of New Zealand's total greenhouse gases emissions (New Zealand's Energy Mix 2021). In August 2019, the government consulted the public regarding the reduction in agricultural emissions and following this meeting it decided to put a price on agricultural emissions from 2025 (Ministry for the Environment 2021b). However, the exclusion of methane from the Zero Carbon Act creates a problem in achieving the emission reduction targets of the country as it is 50% of the country's inventory. Thus, methane emission reductions from agriculture will be much slower than the reduction in emissions of other greenhouse gases (Timperley 2020). In June 2020, the government even after reforming the NZ ETS continued to exempt the agricultural sector from the price of emissions (Climate Action Tracker 2021a). Although the country claims that these domestic targets will help New Zealand in the long term to transit smoothly to a low-emissions future, however, there are doubts regarding their execution.

According to Climate Action Tracker, the actions taken up by New Zealand to mitigate climate change are highly insufficient. Though it appreciates the introduction of a Zero Carbon Act in its law, however, the flawed structure of the act is the problem they are most concerned about (Climate Action Tracker 2021b). Exclusion and setting up of a different target for methane will only increase the problem as the country relies on the mitigation potential of land use and forestry sector instead of the high emitting sectors. Historically also, New Zealand has relied on its forestry sector acting as a carbon sink. The sector offset around 40% of emissions from other sectors between 1990 and 2017. The new act will help in continuing this practice (Timperley 2020). Thus, New Zealand will keep on hiding its rising emissions behind its accounting approach leading and in 2017 emissions have increased by 23% since 1990. Thus, the country will not reduce all its emissions by 2050 as it continues to exempt agriculture and waste sector emissions of methane.

The passing of a zero climate framework would not do the work unless New Zealand implements some strict policies that help in cutting emissions. Moreover, there are doubts on how the government will assess the net-zero targets along with transparency problems. The legal architecture would only provide support but some hard policy measures need to be adopted.

How, Why, and Who Impact These Policies?

Even though climate change has come to be acknowledged as an issue that needs immediate action, the climate policies of Australia and New Zealand seem to be insignificant in tackling it. The absence of proper implementation and strong measures may be cited as one of the reasons. However, the pressure from various interest groups and even the government's concerns regarding the economy of the country play an important role. The economic and political environment of a country plays a major role in the introduction or absence of certain policies in any country. In this section, we will look at the factors that influence the decision-making and hence the policies of both countries. We will then go on to examine the role of various interest groups in the politics of policy outcomes both at the domestic and international level.

In the case of Australia, there will be two types of economic costs that the country will be facing if it accepts mandatory reductions in its greenhouse gas emission. These costs have been explained by Hamilton in his paper titled, "Climate Change Policies in Australia". The first cost is that the country needs to reduce its emissions to the levels it agreed and that too within the predefined period. The second cost that Australia will face is regarding other countries. That is to meet the obligations under the FCCC the countries will be reducing their emissions. As other countries reduce their emissions, they will shift away from coal and towards natural gas, energy efficiency, and renewable. As the worldwide demand for coal slows down and probably declines, Australia's trade with the rest of the world will be affected significantly and it might also decline (Hamilton 2000: 53–54). He is true in assessing these costs as any country that deals with greenhouse gases in its energy and industrial sector will face these costs at any point in time if they seek to reduce their emissions.

The case of New Zealand also revolves around the economic interests of the country being paid more attention to than the environment. As Kelly, puts in her paper, the perception of the public interest was constructed in a quite narrow perspective of national economic interest. That in turn was influenced by a well institutionalized political frame of thought embodying concepts of individualism, small government, economic rationalism, and a commitment to markets as an organizing principle. She further notes that the use of market-interventionist instruments itself contravened the orthodoxy of the day, with its emphasis on the primacy of markets. Thus, noting that any effective policy instruments had to influence emitters in a manner as to reduce their emissions. That was unlikely to be acceptable to the Business Roundtable, to farmers, or even to the general public as the principal generator of transport emissions (Kelly 2010).

The economic benefits that the country would derive from trade are given more importance in both the countries rather than the costs associated with not dealing with climate change problem. Hence, the economic costs of taking action in both countries far outweigh the economic costs of not taking action. This is, however, a narrowly perceived view. As put in the book "The Madhouse Effect", when economists consider the cost of inaction and weigh it against taking action on climate change, the act of reducing carbon emissions by way of taking action is a "no-brainer". That is they say that the cost of not taking action is weighed greater than the cost of taking action because not taking action will cost you in future and this cost will keep on increasing over time (Mann and Toles 2016). Therefore, the cost of not taking any action and citing economic interest as one of the reasons for the same; is making the future possibility of taking action more difficult as the economic costs of reducing emission and taking action would be larger.

Even after knowing the economic costs of not taking action now, the public policy on climate is paralysed. The benefits from the economics of not taking action cannot be separated from the politics of influencing of policies. Thus, it becomes important to understand the players who affect the policy formation and responses. Some people and organizations affect the government's decisions in any country. These are called interest groups as they have their interests in certain departments, policies, and also the outcome of the measures taken by the government. The climate policies adopted by the government of any country impact and also get impacted by these interest groups. Sometimes this happens in the form of stronger policy formation while sometimes it often leads to denial of certain policy measures. Policy and politics enter the stage once a perceived environmental threat has been identified. The disputes are mediated and settled mainly through the states to achieve desired outcomes as interest groups raise their concerns to the government (Bulkeley 2000: 34). The politics of policymaking lead us to think about the groups that affect these policies, both in positive and negative ways. It also leads us to examine the vulnerability and denial of these policies.

These interest groups work at domestic, national, and international levels and have considerable influence on the policies and politics of any country. The domestic interest groups affect the policies first in any country. These interest groups with the help of media and others also run the campaign of deliberate misinformation. As Mann puts it, once it involves the public battle over policy-relevant science, special interests have recognized for a long time that they enjoy certain privileges as the court of public opinion gives preference to their interests. Only a little uncertainty about the scientific evidence in the public mind can ensure that no significant action is taken. They do this through their internal research and focus groups. He further notes that the success of the industry-funded climate change—denial machine derives in part from media outlets' willingness to emphasize conflict over consensus, controversy over comprehension (Mann and Toles 2016). In this way, these groups become successful in delaying action. Thus, there is a need to educate the public about the issue of climate change so that they can become stakeholders in the decision-making process. Moreover, a lot can be learned from how different countries deal successfully with the issue of climate change.

The international pressure, created through both economic and political ways, also influence the policy formation in any country. In economic terms, the introduction of carbon border tariffs or Carbon Border Adjustment Mechanism (CABM), recently by European Union (EU), as they seek to re-level the climate action playing field, increases pressure on the coal and other fossil fuel exporting countries such as Australia and New Zealand. This can be understood simply as, the introduction of the tariff leads to an increase in the price of imports from these greenhouse gas emitting countries thus leading to a decrease in demand for these products. As the export demand falls, the national gross domestic product (GDP) of the fossil fuel exporting country falls with subsequent impact on other sectors of the economy. However, this just gives a simple picture of the more complex mechanisms at play. Thus, by imposing carbon border tariffs, these countries create pressure on the polluting industries to opt for more renewable sources of energy so as to avoid the loss in their GDPs.

According to a survey, only 82% of Australians believe that climate change is real and it is not difficult for the Australian government to change its climate policies (Smith-Schoenwalder 2021). However, international pressure might affect Australia's climate policy. This is where the political aspect of international pressure comes into play. The UN has criticized Australia for having low targets for reductions in its emissions. Thus, Australia is now facing challenges at the global level. Though most of the political aspects of international pressure are not much pronounced, however, Australia's Prime Minister Scott Morrison, in the coming years, would have to deal with international pressure, especially the pressure from the US President Joe Biden to take action towards climate change. New Zealand on the other hand will keep facing international pressure as it has some responsibility towards the Maori ethnic group. Moreover, the Pacific island of Tokelau, administered by New Zealand, also adds a bit of a twist to the policy formation on climate change in the country. Thus, any loss or damage to the area of cultural habitation of these groups and regions due to the ineffective climate response of New Zealand will result in the country coming under excessive international pressure.

While the pressure from the international organizations did have a significant role to play in influencing the climate change responses of the two countries. Some voices say that when comparing the international and domestic pressures, it is the domestic pressures that dominate and make the government work towards policy framing. Carey of CSIS says that negotiations from other countries on climate efforts could make some difference but that domestic pressure could have a larger impact. "Really countries have to figure this out each on their own in terms of their local politics, and G20 and [United Nations] pressure might be able to make a bit of a difference at the margin, but it's really not the driving force in these actions, unfortunately", Carey says (Smith-Schoenwalder 2021). Thus, although considered not so important relatively, however, the political aspect of international pressure cannot be ignored. Hence, both economic and political aspect of international pressure does have a role in influencing the climate policies of the two countries.

SIMILARITIES AND DIFFERENCES IN THE CLIMATE POLICIES OF AUSTRALIA AND NEW ZEALAND

Australia and New Zealand have adopted various climate policies since the initial years of the recognition of the problem and the actions taken for the same. While the dealing of Australia with climate change starts in the 1980s that of New Zealand began in 1972 with the world's first national "green" party called the Values party. After the establishment of the Intergovernmental Panel on Climate Change (IPCC) in 1988. New Zealand was quick in taking steps towards tackling climate change as compared to Australia even though the number of researchers and research organizations in the latter were more compared to the former. New Zealand Climate Change Programme (NZCCP) was established in 1988 and it was coordinated by the Ministry for the Environment. In Australia, National Greenhouse Advisory Council (NGAC) came up in 1989 followed by which the country adopted the Interim Planning Target in 1990. Thus, both the countries were proactive during the initial years.

Both the countries respect the Kyoto Protocol and are trying to develop strategies to reduce their carbon emissions. Post Kyoto, while Australia has focused its strategies on pricing and trading its carbon emissions, New Zealand has recently announced its plan of the Emission Trading Scheme and the introduction of its first Emission Budget from 2022 onwards. Australia has also upgraded its previous National Climate Adaptation Framework 2007 to Climate Change Adoption Programme. While New Zealand is still in the process of the implementation of its National Adaptation Plan and is also working towards an Emission Reduction Plan. The Carbon Neutral Government Programme is introduced in New Zealand recently. No similar step has been taken towards the carbon-neutral scheme in Australia.

The similarities and differences in both countries' climate policies are many and one can trace them easily. Australia even though acquiring a bigger landscape than New Zealand started its journey later than New Zealand in taking action towards climate change. Moreover, the attitude of Australia reflects its non-commitment towards climate policy and action in comparison with New Zealand. The successful passing of the Zero Carbon Act in 2019 in the New Zealand legislature reflects the countries commitment at least on paper towards the efforts regarding emission reductions. Another noticeable feature is that the act was passed with broad party consensus. This may help the country to be committed to taking action on climate change as the successive governments in New Zealand come to power.

The non-commitment of Australia is reflected in part by not taking as significant measures as New Zealand has taken, as the introduction of emission reductions and action on climate change in its law. And in part by how the country has disbanded its Department of Climate Change and Energy Efficiency, the independent department that formulated the country's climate policies earlier. The climate policies of the country are now framed under the Department of Industry, Science, Energy, and Resources. This reflects either the government's lack of political will to take action on climate change or the increasing influence of industrial groups on their decision-making or both. The country is the world's largest exporter of coal, iron ore, uranium, and natural gas. The government backing of these industry groups is also happening because of the trade and other economic benefits that the economy derives through them. The pressure from industrial groups is huge in the country as Australia is dependent on greenhouse emission-intensive industries.

The industrial pressure in New Zealand is also significant especially concerning how the emission-intensive industries are freely given their New Zealand Emission Requirements (NZU) to support the trade and the economy. Even the exclusion of biogenic methane from the Zero Carbon Act 2019 reflects how the country is neglecting the emissions from the sector which constitute about half of the total emissions in the country.

Thus, in both countries, the policies for action against climate change are formulated keeping in mind the economic considerations. In addition, the pressure from various stakeholders influences the policies largely. However, one feature that comes out from this is that in Australia the population is somewhat hostile towards the idea of climate change. While in New Zealand, the introduction of the Zero Carbon Act (2019) in its law showcase the concerned attitude of the population of the country towards the issue of climate change. Therefore one can trace not only the similarities and differences in the policies of the two countries but also how and who affects the policies' vulnerability and sometimes their denial.

Conclusion

Climate change is a real thing. So is climate denial. The Pacific is experiencing climate change. The countries in the region are experiencing it already. Australia and New Zealand are two prime examples. The problem of climate change is evident more so now. However, with the adoption of adequate and proper policy measures, this problem can be solved. Both, Australia and New Zealand, have taken steps for the same. There are many similarities and differences in their policy towards action on climate change. Although both of these countries' engagement on the issue spans more than three decades now. Still, the efforts of Australia and New Zealand are regarded as highly insufficient in tackling climate change. Various interest groups are affecting the decision-making processes and hence the policy formulation in both of these countries.

The groups that have affected the decision-making and policy formulation are also quite similar, yet different in many ways. These groups work both at the domestic and international levels and have a considerable impact both in terms of economic and political aspects. At the domestic level, the preoccupancy of both of the countries' governments with the economic concerns of the country leads to their lack of political will in the implementation of already formulated policies and strategies. Thus, any new policy development is extremely weakened by the interplay of these forces.

However, this weakening of policymaking can be effectively cured through equal pressure from groups that are concerned about the environment. These groups can impact their decisions and policies on climate by working from international as well as domestic forums. Not just the international organizations but developed and large countries, in particular, could create pressure on them, both through economic and political means. The economic pressure would imply putting restrictions on these carbon-emitting countries. These restrictions can be in the form of tariffs on their carbon emissions. The political pressure would consist of calling them out for their irresponsible acts of greenhouse gas emissions at climate change forums and organizations.

As both these countries are part of the Pacific, they do have some responsibility towards the small and developing island countries in the Pacific region who have much to lose from climate change. Therefore, Australia and New Zealand would not only be confronted with the question of other living beings or the future generations of their own nations but would also have obligations towards the living beings residing in these small island countries in the Pacific. Thus, generating a kind of international pressure from the countries lying in the Pacific region.

However, the domestic pressure, that is the pressure from within these countries will have a greater impact. This domestic pressure will eventually decide the adoption or the denial of the policy. The citizens, civic groups, activists, students, and institutions of these countries will therefore have a big role to play in tackling climate change and its effects through effective policy solutions. They will be the ones who need to keep a check on the adoption and execution of these policies. As the problem of climate change cannot be ignored for long now, there is a need for concerted effort to tackle climate change through positive and effective policies of the two big countries in the Pacific.

References

Antonich, B. (2020). *New Zealand's Updated NDC Informs of Its Zero Carbon Amendment Act*. Available at: https://sdg.iisd.org/news/new-zealands-updated-ndc-informs-of-its-zero-carbon-amendment-act/.

Ardern, J., Nash, S., & Shaw, J. (2020). *Public Sector to Be Carbon Neutral by 2025*. Available at: https://www.beehive.govt.nz/release/public-sector-be-carbon-neutral-2025.

Basher, Reid E. (2000). 'The Impacts of Climate Change on New Zealand', in Alexander Gillespie & Wil Burns eds., *Climate Change in the South Pacific: Impacts and Responses in Australia, New Zealand, and Small Island States*.

Beeson, Mark & Mc Donald, Matt. (2013). 'The Politics of Climate Change in Australia', *Australian Journal of Politics and History*, 59(3), pp. 331–348. Available at: https://www.researchgate.net/publication/261532495_The_Politics_of_Climate_Change_in_Australia/citations.

Bulkeley, Harriet. (2000). 'The Formation of Australian Climate Change Policy: 1985–199', in Alexander Gillespie & Wil Burns eds., *Climate Change in the South Pacific: Impacts and Responses in Australia, New Zealand, and Small Island States*.

CCPI. (2020). Available at: https://newclimate.org/wp-content/uploads/2019/12/CCPI-2020-Results_Web_Version.pdf.

CCPI. (2021). Available at: https://ccpi.org/country/aus/.

Climate Action Tracker. (2021a). Available at: https://climateactiontracker.org/countries/new-zealand/policies-action/.

Climate Action Tracker. (2021b). Available at: https://climateactiontracker.org/countries/new-zealand/.

Climate Change Commission. (2020). Available at: https://www.climatecommi ssion.govt.nz/who-we-are/our-story/.

Daalder, M. (2021). *Overseas Expert Criticises Climate Commission's Accounting.* Retrieved August 17, 2021, from Newsroom website: https://www.new sroom.co.nz/overseas-expert-hits-climate-commissions-accounting.

DCCEF, Department of Climate Change and Energy Efficiency Tackling the Challenge of Climate Change (Canberra, 2012).

Department of Environment, Sport and Territories. (1995). *Greenhouse 21C: A Plan of Action for a Sustainable Future.*

DISER. (2021). Department of Industry, Sciences, Energy and Resources. Available at: https://www.industry.gov.au/policies-and-initiatives/australias-climate-change-strategies.

Downes, D. (1996). 'Neo-Corporatism and Environmental Policy', *Australian Journal of Political Science*, 31(2), pp. 175–190.

Goodman, Jack. (2020). 'What Is Australia Doing to Tackle Climate Change', BBC, January 2, 2020, Available at https://www.bbc.com/news/world-aus tralia-50869565

Hamilton, Clive. (2000). 'Climate Change Policies in Australia', in Alexander Gillespie & Wil Burns eds., *Climate Change in the South Pacific: Impacts and Responses in Australia, New Zealand, and Small Island States.*

Industry Commission Annual Report, 1989–90. (1990). Available at: https:// www.pc.gov.au/research/supporting/industry-commission-annual-report-1989-90/industry-commission-annual-report-1989-1990.pdf.

International Climate Change Commitments. (2021). Available at: https:// www.industry.gov.au/policies-and-initiatives/australias-climate-change-strate gies/international-climate-change-commitments.

International Energy Agency. (2017). Available at: https://www.iea.org/pol icies/481-the-greenhouse-challenge-challenge-plus-industry-partnerships.

International Energy Agency. (2018). Available at: https://www.iea.org/rep orts/coal-2018.

International Energy Agency. (2021). Available at: https://www.iea.org/art icles/new-zealand-climate-resilience-policy-indicator.

IPCC. (2021). Available at: https://www.ipcc.ch/about/history/.

Johnston, Tim. (2007). 'For Rudd, Hard Part Comes after KYOTO', *New York Times*, December 4. Available at: https://www.nytimes.com/2007/12/04/ world/asia/04iht-australia.1.8580545.html

Kelly, Geoff. (2010). Climate Change Policy: Actions and Barriers in New Zealand. Sydney Business School - Papers.

Kinrade, P. (1995). 'Towards Ecological Sustainable Development: The Role and Shortcomings of Markets', in R. Eckersley ed., *Markets, The State and The Environment: Towards an Integration.*

Leining, Kerr, & Bruce-Brand, Bronwyn. (2020). 'The New Zealand Emissions Trading Scheme: Critical Review and Future Outlook for Three Design Innovations', *Climate Policy*, 20(2), pp. 246–264. https://doi.org/10.1080/146 93062.2019.1699773.

Mann, M. E., & Toles, T. (2016). *The Madhouse Effect: How Climate Change Denial Is Threatening Our Planet, Destroying Our Politics, and Driving Us Crazy.*

Ministry for the Environment. (2006). Available at. https://environment.govt. nz/assets/Publications/Files/4th-national-communication-2006.pdf

Ministry for the Environment. (2019). Available at: https://environment. govt.nz/acts-and-regulations/acts/climate-change-response-amendment-act-2019/.

Ministry for the Environment. (2021a). Available at https://environment.govt. nz/what-government-is-doing/areas-of-work/climate-change/emissions-red uction-targets/greenhouse-gas-emissions-targets-and-reporting/.

Ministry for the Environment. (2021b). Available at: https://environment.govt. nz/what-government-is-doing/areas-of-work/climate-change/about-new-zealands-climate-change-programme/.

National Climate Resilience and Adaptation Strategy. (2015). Australian Government. Available at: https://www.environment.gov.au/system/files/resour ces/3b44e21e-2a78-4809-87c7-a1386e350c29/files/national-climate-resili ence-and-adaptation-strategy.pdf.

National Communication on Climate Change (Australia). (2013). A Report Under the UNFCC. Available at: https://unfccc.int/resource/docs/natc/ aus_nc6.pdf.

NBC. (2007). *Australia Ratifies Kyoto Global Warming Treaty.* https://www. nbcnews.com/id/wbna22081582.

New Zealand's Energy Mix. (2021). Available at: https://www.energymix.co. nz/our-alternatives/new-zealands-emissions/.

Nogrady, Bianca. (2019). 'Scientists Criticize Australis's Questionable Climate Policy', *Nature*, February 28. Available at: https://www.nature.com/articles/ d41586-019-00725-6.

Scott, Jason, & Jess, Shankleman. (2020). 'Australia Excluded from Global Climate Talks as Policies Lag', *Bloomberg*, December 10. Available at: https://www.bloomberg.com/news/articles/2020-12-10/australia-excluded-from-global-climate-meeting-as-policies-lag.

Smith-Schoenwalder, Cecelia. (2021). 'Australia Faces Uphill Battle to Address Climate Change', April 21. U.S.News. Available at: https://www.usnews. com/news/best-countries/articles/2021-04-21/australia-faces-uphill-battle-to-address-climate-change

Talberg, Anita, Hui, Simeon & Loynes, Kate. (2016). 'Australian Climate Change Policy to 2015: A Chronology', in Science, Technology, Environment and Resources Section, Research Papers 2015–16, Parliament of Australia. Available at: https://www.aph.gov.au/About_Parliament/Parliamentary_Departments/Parliamentary_Library/pubs/rp/rp1516/Climate2015#:~:text=In%20December%202015%20a%20Liberal%20Government%20announced%20that,meet%2C%20and%20surpass%2C%20its%202020%20emission%20reduction%20target.

Taplin, R. (1996). 'Climate Science and Politics: The Road to Rio and Beyond', in A. Henderson-Sellers & T. Giambelluca eds., *Climate Change: Developing Southern Hemisphere Perspectives.*

Taylor, Rob. (2007). 'Australia's Rudd Will Sign Kyoto Pact If Wins Vote', Reuters. Available at https://www.reuters.com/article/us-australia-election-idUSSYD372620071119.

Taylor, Phil. (2020). 'New Zealand Declares a Climate Change Emergency', *The Guardian*, December, 2. Available at: https://www.theguardian.com/world/2020/dec/02/new-zealand-declares-a-climate-change-emergency.

Timperley, Jocelyn. (2020). 'What Can the World Learn from New Zealand on Climate?' *The Lancet Planetary Health*, 4(5), pp. 176–177. ISSN https://www.sciencedirect.com/science/article/pii/S2542519620301091.

INDEX

Printed by Printforce, the Netherlands